William Battle Phillips

Iron making in Alabama

William Battle Phillips
Iron making in Alabama
ISBN/EAN: 9783743398153
Manufactured in Europe, USA, Canada, Australia, Japa
Cover: Foto ©berggeist007 / pixelio.de

Manufactured and distributed by brebook publishing software (www.brebook.com)

William Battle Phillips

Iron making in Alabama

EUGENE ALLEN SMITH, PH. D., DIRECTOR.

IRON MAKING

IN

ALABAMA,

BY

WILLIAM BATTLE PHILLIPS, PH. D.,

Consulting Chemist Tennessee Coal, Iron & Railroad Co.,
Birmingham, Ala.

MONTGOMERY, ALA., 1896:
JAS. P. ARMSTRONG, PRINTER.

ERRATA.

P. 1, line 4, from bottom—Peckin should be Pechin.
P. 1, line 9, from bottom—Erskin should be Erskine.
P. 15, line 13, from bottom—Ore should be Ores.
P. 41, lines 5 *et seq.* from top—Silica x alumina should be silica plus alumina.
P. 57, line 8 from bottom—61 should be 61.2.
P. 57, line 9 from bottom—68.8 should be 58.8.
P. 64, line 3 from top—Uchling should be Uehling.
P. 120, line 19 from bottom—Annisto should be Anniston.

To His Excellency,
> WILLIAM C. OATES,
>> *Governor of Alabama :*

DEAR SIR :—I have the honor to transmit herewith a Report upon Iron Making in Alabama, by Dr. Wm. B. Phillips.

A glance at the Table of Contents will show how completely the ground, from the raw materials to the finished product, has been covered by the author, and the reader of the book will soon perceive that the various topics have been more fully and carefully treated than ever before.

This report will be an invaluable, and at the same time authoritative, handbook of all the conditions which surround the iron-making business in Alabama, and as such, is certain of a hearty welcome, not only from our own citizens, but from all others interested.

> Very respectfully,
>> EUGENE A. SMITH,
>>> State Geologist.

University of Alabama, July 1st, 1896.

	Pages.
Letter of Transmittal....	1–2
Introduction	3–11

CHAPTER I.

The Ores—General Discussion—Kinds used—Bessemer ore not found in quantity—Phosphorus in ores—Value of in State—But little bought on analysis—Evils of purchasing by ton—Improvement of ores—Pig iron made almost exclusively from local ores—Production and value—Rank of the State as an ore producer—Transportation of pig iron to market the main question.. 13–28

CHAPTER II.

The hematite ores—Classification—Occurrence—The soft red ores—Analysis—Physical nature—Former practice almost restricted to use of soft red—Exhaustion—Concentration—The limey, or hard red ores—Description—Analysis—Crushed—Calcination—The limonite, or brown ores—Occurrence — Mining — Washing — Analysis — Calcining—Valuation—Screening—Mill Cinder—Analysis—Blue Billy—Purple ore 29–56

CHAPTER III.

The fluxes—Limestone—Dolomite—Analysis—Valuation—Sold on analysis—Dolomite as flux compared with limestone 57–67

CHAPTER IV.

The fuels—Coke—Classification—Analysis—Cell space—Specific gravity—Crushing strain—Statistics of production—Character of coal used—Composition of ash............... 68–77

CHAPTER V.

Furnace burdens—Coke—burdens of hard and soft red—Consumption of raw materials—Cost of raw materials—Deductions—Burdens of hard and soft red, and brown ore—Deductions—Charcoal furnaces—Burdens—Cost of raw materials................ 78–105

CHAPTER VI.

List of Furnaces, Rolling Mills, &c. in Alabama—Production of coke and charcoal iron—Hot blast stoves—Progress of furnace building 106–125

CHAPTER VII.

Pig iron market—Grading of coke pig iron—Freight tariff—Prices—Freight tariff for coal and coke—Production of coal, coke and pig iron........ 126–159

LETTER OF TRANSMITTAL.

Dr. Eugene A. Smith,
  Director Ala. Geol. Survey,
    University, Ala.

SIR—I beg to transmit herewith a report on Iron Making in Alabama, prepared for the Geological Survey.

No systematic attempt has yet been made to bring this industry to the attention of the general public. Numerous articles have appeared in the technical papers in this and other countries during the last ten years, dealing with special phases of the subject, and many of them possess great merit. In particular may be mentioned the following:

The Iron Ores and Coals of Alabama, Georgia and Tennessee. Jno. B. Porter, Trans. Amer. Inst. Min. Engrs., vol. xv, 1886–87, pp. 170–2'8.

Comparison of Some Southern Cokes and Iron Ores. A. S. McCreath and E. V. D'Invilliers. Trans. Amer. Inst. Min. Engrs., Vol. xv, 1886–87, pp. 734–756.

General Description of the Ores used in the Chattanooga District. H. S. Fleming. Trans. Amer. Inst. Min. Engrs., Vol. xv, 1886–1887, pp. 757–761.

The Pratt Mines of the Tennessee Coal, Iron and R'y Co. Erskin Ramsay. Trans. Amer. Inst. Min. Engrs., Vol. xix, 1890-91, pp. 296–313.

Notes on the Magnetization and Concentration of Iron Ore. Wm. B. Phillips, Trans. Amer. Inst. Min. Engrs., Vol. xxv, 1895–1896.

A series of articles by E. C. Peckin, in the Iron Trade Review in 1888, and by the same author in the Eng. & Mining Journal, Vol. lviii, 1894. Also the Proc. Ala. Indust. & Sci. Soc. 1891–1896.

But the very fact of their appearing in technical publications has caused the general reader to neglect them, not on account of indifference, but because they were not readily accessible. The files of the great industrial journals, and the Proceedings of the Amer. Inst. of Mining Engineers are not available to many who wish to know what has been already done in Alabama, and what the future may confidently be expected to unfold.

After careful consideration, it was decided to prepare a little book of 150–200 pages which should present the matter as it is to-day and chiefly from the standpoint of raw materials. Very little has been said as to furnace practice, because it was not in mind to prepare a Text-book of Iron Making. The book is intended for general distribution by the Geological Survey, and while the main purpose is to supply the average reader with easily digestible information, it is hoped that those who are actively engaged in the business may find in it some suggestions not altogether unworthy of their attention.

Very truly yours,

WM. B. PHILLIPS.

Birmingham, Ala., May, 1896.

IRON MAKING IN ALABAMA,

BY

WILLIAM B. PHILLIPS.

INTRODUCTION.

During the last twenty-five years so great an improvement in the manufacture of pig iron and its utilization in more or less finished products has taken place in Alabama that it is now thought expedient to describe, as briefly as possible, the conditions that have compassed the industry and that are still in force.

In 1872, Alabama produced 11,171 tons of pig iron; in 1892, 915,296 tons. In 1880, the state produced 60,781 tons of coke, and in 1892 1,501,571 tons. In the census year of 1870 the amount of capital invested in the iron business, including mining, was $605,700, and excluding mining $566,100. In that year the total production of pig iron was 6,250 tons, valued at $210,258, and there were used 11,350 tons of ore valued at $30,175. In the census year of 1890 the capital invested in the mining of iron ore alone was $5,244,906, the amount of ore mined and used being 1,570,319 tons, valued at $1,511,611.

The Southern States generally sell their entire iron product for purposes other than steel making. The iron goes to foundries, mills and pipe works. It was not until recently that any considerable amount found its way into steel works. It is not probable that more than one-twentieth of the iron made in the South goes to the steel maker. Alabama offers no exception to this rule. It was not until the last few months that any fairly large shipments of iron made here were sent to steel plants. The significance of this statement will appear when it is remembered that the total amount of iron produced in the United States in 1895, not intended for steel making, was about 3,000,000 tons. At the present time Alabama is producing 35% of the iron used in the foundries, mills and pipe works of the country. The growth of the industry has been conditioned chiefly by three great factors :

First, the cheapness with which the ores can be mined and delivered.

Second, the proximity of the ore to the flux and fuel.

Third, the tendency of pig iron consumption towards the interior.

The cheapness of an ore is not always to be measured by its cost at the furnace. There are also to be considered its quality in respect of its content of metallic iron and the presence of ingredients which determine the use to which the pig iron made from it can be put. The lower the percentage of iron in an ore the cheaper must it be mined and transported in order that a market for the pig iron may be secured and held. A very rich ore may allow of mining and transportation costs that would prevent the use of an ore less rich. The same principle applies to the quality of the ore as regards its freedom from injurious substances. If it is free from phosphorus and sulphur, for instance, it may be highly acceptable to the steel plants. If at the same time it be rich in iron we may have the conditions that allow of maximum cost at the furnace. In Alabama we have ores of a moderate content of iron, and they must therefore be mined at a low cost. They also contain too much phosphorus to allow of the pig iron being used for making Bessemer steel.

The principle on which the makers of pig iron in Alabama have had to proceed is the utilization of local ores, and the production of suitable coke from native coal. It all seems plain sailing to us now that the yearly output of coke exceeds one and a half million tons, and the yield of pig iron is above 800,000 tons; but twenty years ago it was by no means certain that good coke could be made from Alabama coal on a large scale, and the use of Red Mountain ores was a vexed question. As late as 1883, so-called representative analyses of Alabama hematite were published showing 56% and 61% of iron on the one hand, while on the other it was said that pig iron made from Alabama ore and coke was so brittle that it ought to be kept under glass as a curiosity. Both these statements were equally removed from the truth. When finally it became known that with but few exceptions the Red Mountain ores could not be expected to contain more than 47% of iron as mined and that the fifty-six and sixty-one per cent. hematite ores could be exhausted in a single day, the situation rapidly improved. So far as the ores were concerned, the problem narrowed down to the single question whether they could be successfully used in conjunction with cokes of domestic production. From that day to the present the question has changed but little, the main difference being that the price of ore has steadily diminished, reaching its lowest point in 1895, and that the coke is better and cheaper. During a part of this year the price of soft red ore, analyzing about 46 per cent. of iron, was fifty cents per ton, stock house delivery. It was during this year also that the cost of making pig iron in Alabama was at the lowest, less than $6 per ton. No more striking illustration of the great change that has come over the manufacture of pig iron in Alabama during the last few years can be adduced than to say that the total cost of production is now less than the cost of the raw materials five years ago. This has been rendered possible not only by reductions in the cost of the raw materials, but also and particularly by improvements in furnace practice and a closer alliance between the chemist and the superintendent. There is a large iron company in the state which three years ago had no chemist, and the laboratory which had formerly been tenanted had been allowed to take care of itself for two years. This company has now four chemists in its employ and one of the best equipped laboratories in the country. Three years ago it was content to have some of its materials analyzed perhaps once a month; now the number of analyses per month is close upon four hundred. Chemical inspection of the stock goes hand in hand with inspection of the product, and there is now not a single thing used or made whose composition is not known. A great amount of material is bought and sold on analysis, and the inevitable tendency is towards the extension of this system to all materials. The most progressive companies in the state are now recognizing the value of close chemical inspection of the ores, fluxes and fuels. In this respect the change that has come over the industry during the last five years is particularly noticeable and must be regarded as one of the most hopeful signs of the time.

Another agreeable improvement in the business is the willingness of the iron masters to exchange information and opinions, to visit competitive establishments and cultivate the more social side of trade. There need not be rankling jealousies between those engaged in similar enterprises in the same district. To refuse to impart information is to refuse to acquire it, and the day has long since passed when in the mind of any one man is to be sought correct knowledge on all phases of the same matter. Without such cordial interest in what may be for the general good, this sketch of the materials used in making iron in this state, however imperfect it may be and doubtless is, could not have been undertaken in any hope of success. My own acquaintance with the district dates from 1887, and since that time I have accumulated nearly 10,000 analyses of every kind of material used in making iron in the state, coming partly from my own laboratory and partly from the records of companies actively engaged in the production of iron. The deductions that will be met with in the body of this Report are founded upon analyses that were made in the interest of those prosecuting the iron business, not upon analyses of stray fragments or hand specimens. They represent hundred of thousands of tons of ore, limestone, dolomite, coal and coke, the samples being drawn from the stockhouses during a period extending over many years. In numerous instances samples of the ore were taken direct from the mines, foot by foot down the seam, and from mine and railroad cars. The constant effort has been not to include in the pages of this report any conclusions that were not based upon the actual practice in the State and District, and the reader is assured that no pains have been spared to accomplish this end.

To those who have most generously given the information desired of them, I would express my hearty thanks. It is a source of great pleasure to me that the replies to requests of this nature should have been met so fully and so courteously, and I trust that the interest in what the state has to offer to the makers of iron may be deepened and broadened from this attempt to set in order the results already attained.

According to Swank (History of Iron in all Ages, 2nd Ed., p. 293, et seq.), who quotes from Leslie, the oldest furnace in Alabama was built about 1818. It was a charcoal furnace, and was situated a few miles west of Russellville, Franklin county, doubtless to use the brown ore of the Russellville belt, which is of excellent quality and is now used by the coke furnaces at Sheffield. It seems to have been abandoned about 1827, and from that date until 1888, a period of 60 years this deposit of ore remained undeveloped and unused. Not long since there came to hand evidence of the existence of this old furnace in the shape of a piece of very impure iron which was brought to the writer from that part of Franklin county by a person who supposed it was iron ore.

From 1827 until 1843, there is no record of any furnace building in the State, the next one being at Polksville, Calhoun county; then one at Shelby, Shelby county, in 1848 ; and one at Round Mountain in 1853.

Charcoal iron has been made at Shelby almost continuously since 1848, and the reputation of the iron has not been excelled from that day to the present time.

The furnace was built by Horace Ware, who afterwards added a foundry and a mill for cotton ties and bar iron. This furnace was burned in 1858, but rebuilt at once. A larger mill was built in 1859, and iron rolled April 11th, 1860. This mill was very active during the war of the Confederacy, and was burned by the Union troops under General Wilson in 1865. It has not been rebuilt, but a part of the machinery was used in constructing the rolling mill at Helena in 1872. It may not be amiss at this point, while briefly considering this historic furnace and mill to quote a very interesting letter written by Mr. E. T. Witherby, assistant secretary of the Shelby Iron Company to Mr. Swank in 1888. "The first blast furnace erected here went into blast in 1848. Horace Ware was its proprietor. In 1854, Mr. Robert Thomas made iron in a forge near here. This iron was sent to England and returned in razors and knives. In 1859 Mr. Ware began the erection of a rolling mill. It was completed and started in the spring of 1860. In 1862 Mr. Ware sold his property to the Shelby County Iron Manufacturing Company, which erected a new furnace, the one which we have recently torn down, and on whose site we are erecting a new stack. The rolling mill was enlarged in 1862, and was operated continuously until March 31st, 1865, when it was destroyed by General Wilson of the Union army. It was in this mill, in 1864, that the plates were rolled for the armor of the iron clad ram *Tennessee*. Judge James W. Lapsley, one of the stockholders and directors of the present Shelby Iron Company, was made a prisoner by the Union forces in 1863, while in Kentucky looking for puddlers for this mill.

When I came here, nearly twenty years ago, we had plates, merchant bars, and strap rails on hand made entirely of Shelby iron and rolled in this mill. Some of the plates, known to us now as the "gun boat iron" are still in our store house, but they have been slowly disappearing under the demand of our blacksmiths for "an extra good piece of iron" for "this job," or that "particular place." Some of these plates are 8 inches by 3 inches, and others 11 inches by 5 inches, and of various lengths; originally, they were, perhaps, 10 feet long. Shelby pig iron was also shipped to the Confederate arsenal and foundry at Selma, Alabama, in 1861, where the *Tennessee* was constructed and fitted out. This iron doubtless went into guns and other castings for this vessel. Catesby ap Jones was superintendent of the arsenal, and with his senior in rank, Franklin Buchanan, both pupils of that sea-god, Matthew Calbraith Perry, wrought out the *Tennessee*. They were as full of progressive ideas regarding steam and armor as their master, and nothing but the scanty means at their disposal prevented a much more formidable iron-clad than the *Tennessee* from being set afloat."

Car-wheel makers are the exclusive users of our iron.

It is interesting to note in connection with the Confederate States foundry at Selma, that it used coke made from the Gholson seam mined at Thompson's Lower Mine, on Pine Island branch, in Sec. 10, T. 24, R. 10 E., Bibb county, and elsewhere in the vicinity, as we are informed by Eugene A. Smith (Ala. Geol. Survey, Report of Progress for 1875, pp. 32 and 33.) This was about 1863, and is probably the first use of Alabama coke for foundry purposes.

"In 1863-64 Capt. Schultz of the Confederate army made a large quantity of coke from seams in the Coosa coal field, getting it to market by floating it down the river in flats to the railroad bridge across the Coosa river, whence it was carried by rail to Montgomery and Selma. This coke was said to be the finest ever made in the State, and to equal the very best English cokes." (Smith ut supra, p. 38.)

In 1825, there was a bloomary near Montevallo, Shelby county; several in Bibb county in 1830-1840; one in Talladega county in 1842; two in Calhoun county in 1842. In 1856 there were enumerated 17 forges and bloomaries, about one-half being in operation and producing 202 tons of blooms and bar iron. The total product of charcoal pig iron in 1856 was 1,495 gross tons.

In 1876 the Eureka Coke Furnace was built at Oxmoor, Jefferson county, by Col. J. W. Sloss, one of the most active iron-masters in the State, and the founder of the coke iron industry. This was the first furnace to go in on coke, and was followed in 1880 by the Alice furnace, built at Birmingham in 1879-80, by H. F. DeBardeleben, another noted name in the history of the iron trade in Alabama. Then followed the first of the Sloss furnaces at Birmingham, built by Col. J. W. Sloss in 1881-82, and put in blast April 12th, 1882.

Space would fail us to enumerate the names of those concerned in the early history of the coke iron trade in Alabama, but J. W. Sloss (who died in 1890), H. F. DeBardeleben, T. T. Hillman and Geo. L. Morris, who are still enjoying the fruits of their foresight and energy, will always be first called to mind by the historian of the days, not long past as we measure years, but removed from us by a continuous series of splendid achievements.

Si monumentum quæris, circumspice.

WM. B. PHILLIPS.

BIRMINGHAM, ALA., May, 1896.

IRON MAKING IN ALABAMA.

CHAPTER I.

THE ORES: GENERAL DISCUSSION.

The ores used in the production of pig iron in Alabama fall naturally into two classes, and for convenience of reference the local names will be used with full explanations under each. They are either limonites, the so-called "brown ores," or hematites, the so-called soft, and hard ores. There are deposits of blackband ores and of magnetites, none of which, however, come into use. Efforts have been made to use the more or less bituminous blackband ores, both raw and calcined, but they were not successful. Several years ago an attempt was made at one of the coke furnaces to employ the raw black-band ore found in association with one of the coal seams in the northern part of Jefferson county, but the furnace worked badly, probably owing to the very bituminous nature of the ore, and the experiment was discontinued. This same ore was afterwards calcined in piles in the open air and a portion of the resulting material was of fair quality. But owing; it is thought, to the lack of care in the management of the business there was a good deal of trouble from the caking of the ore. In places it resembled impure iron and was almost malleable. Nothing has been done in this direction for some time, as the available supply of ores that do not need such treatment is still very large. Practically all of the iron made in the State has been produced from limonite, hematite, or a mixture of the two.

For special purposes, as for instance, car wheel iron or some particular kind of iron destined for the pipe works, brown ores alone are used, although at times some admixture of hematite is permitted even then. For ordinary foundry and mill irons, and of late for basic iron, the common practice is to use a mixture of brown and hematite ores, the proportion of brown ore being for the most part about 20 per cent. of the ore burden, although there are some important exceptions to this rule.

It seems best to take up the ores under separate headings-that a fuller understanding of the subject may be reached, but before doing so some observations on the ores in general may not be out of place.

In Alabama a vast deal of prospecting has been carried on for more than twenty years to ascertain if it were possible to find richer ores or ores of cheaper accessibility. During the flush times several chemical laboratories were in active operation in more than one town and thousands of analyses were made of almost every known deposit. In many cases the samples were taken by interested persons and in many others by persons wholly unacquainted with the first principles of sampling ore seams. In the writer's own experience it has happened scores of times that a single piece of ore, not larger than the fist, would be brought in as representing the seam. In one case of the kind it happened that the ore showed a comparatively small amount of phosphorus, with some 46 per cent. of iron. Whereupon the report was circulated that a large deposit of Bessemer ore had been discovered and for a while speculators were busy. If there is any large deposit of Bessemer ore in the State it has not yet been found. There are places where some of the brown ores show phosphorus below the Bessemer limit, but fifty feet away they are liable to carry from 0.20 per cent. to 0.50 per cent. of this element. The same observation applies to a certain seam of fine grained soft red hematite. Many seams have been carefully sampled and many analyses made in the search for ore that would not show phosphorus above the Bessemer limit, i. e., not over 0.04 per cent. for 50 per cent. of iron. But the conclusion has finally been reached that for the present we shall have to confine ourselves to ores that contain from 0.10 to 0.40 per cent. of phosphorus per 50 per cent. of iron, and in many of the brown ores we may expect a considerable increase over these figures. It will not be denied that for a small furnace and with great care in the selection of the ore, the chemist being constantly employed in analyzing for phosphorus, it might be possible to make Bessemer iron in this State from some of the brown ores, but no one could be advised to undertake the project with present lights. The attempt has been made and several thousand tons of iron with less than 0.10 per cent. of phosphorus were produced, but the enterprise languished and has not been revived.

The treacherous nature of brown ore with respect to its content of phosphorus, no less than with respect to the continuity of the deposit, is enough to forbid reasonable hope of success.

The hematite ore, on the other hand, carry phosphorus much above the Bessemer limit. They carry generally from 0.30% to 0.40% of phosphorus, although there is in the district contiguous to Birmingham a small seam of red hematite that carries 5.41% of phosphorus and another 2.31%, the metallic iron being about 38%.

In the early days of iron making in the Birmingham district it was the rule, according to one of the contractors, "to mine anything that was red," and what was mined went into the furnace. The difference between good, bad and indifferent may have been known, but was not a factor with the contractor or with the furnace manager. All this has been changed and it is now required that the ore shall be something more than merely red.

The principles underlying the valuation of iron ores are but little used in the state, the old system of purchasing by the ton still being maintained. The average value of the ores in 1894 was given by Mr. John Birkinbine (Mineral Resources of the United States, U. S. Geol. Survey, 1894) at 83 cents per ton, a decline of 13 cents since 1889, Accepting this as correct, Alabama would rank third in the list of states showing a low value per ton of ore, Minnesota being first with 73 cents and Texas next with 75 cents. The value of an ore is the price at the mine, for, unless the minor also pays the freight, he has already added to the cost of mining all the legitimate costs that should apply to a ton, including royalty. If his contract requires that he pay the freight, he can not reasonably add the freight to the value of the ore, for this varies with the distance it has to be transported.

With the exception of some brown ores, which are purchased on the unit basis, but which constitute a small part of the ore used, and some special contracts relating to hematite, the ores in Alabama are bought by the ton without regard to their composition. The price is so much per ton, whether they carry forty, or forty-three, or forty-seven, or fifty per cent. of iron.

This system has but little to recommend it, except a mistaken notion of economy in the saving of laboratory expenses and sampling. A close inspection may be kept on the ore as received and daily reports made as to its composition, but unless there is a penalty attached to the shipping of poor ore, there is really no way in which it can be stopped. The price is uniform, no matter what the ore may be. It may be improperly mined, it may contain unusual amounts of water, or clay, or chert, but the price is the same to the furnace. A car load of

GENERAL DISCUSSION.

ore may contain 47% of iron to-day, to-morrow the ore from the same mine may contain only 43%, yet the price is the same. A brown ore may reach the furnace with its customary 7% of water, to-morrow it may have 13%, yet the ore is sold by the ton and the water is counted as ore.

There are two main results from this system: First, the contractor is not impelled to furnish ore any better than would be accepted. His sole aim is to avoid disputes with the furnace man by sending ore that indeed could be better but still will pass muster. There may arise under this condition of affairs a tendency towards careless mining, and if the line between acceptable ore and bad ore is an arbitrary one, as is frequently the case, there is a temptation to put the shot down a little bit deeper than the line of separation. In the mining of the soft red ores by open cut, the over-burden having been removed, it is practically impossible to distinguish between ore of 46% iron and ore of 40% simply by the eye. The chemist alone can decide the question. It is a fortunate circumstance, in the Birmingham district, that for the most part the contractors are fully alive to the advantages of shipping ore that will cause no dispute. Under the present system it is difficult to see how they could ship better ore than they do. But the system itself is wrong in principle. The administration of it may be as fair to the contractor as to the furnace, but this does not do away with the main objection to it, which is that the same price is paid for ore that is barely usable as for ore that is really good. It can not be denied that this objection is valid and that until it is removed the true principle underlying the valuation of ores can not be put into practice.

The second result from the system of purchasing ore

by the ton and not on analysis is that the furnace man can not know that his ore to-day is of the same composition as it was yesterday and will be to-morrow. The purchase of ore on analysis does not necessarily condition regularity of stock, but it is a long step towards this most desirable end. It is more than probable that under it there would be a tendency towards the higher grades of ore, for these would be more profitable to the contractor than the lower grades.

The irregularity in the stock is one of the most serious obstacles with which the Alabama iron master has to contend, especially when he is using Red Mountain ores. The most untiring vigilance is demanded in order that the entire make of the furnace shall not be injuriously affected. It is of course the fact that a great deal of excellent iron has been made in the State without calling into constant requisition the services of a chemist. But this is no more than saying that many a case of illness has been cured without the care of a regular physician. We venture the assertion that even under the present insufficient system a lower cost account for the making of iron would be shown by the companies employing chemists than by the others. By far the greater amount of iron now made in Alabama is the product of companies with well equipped laboratories, and some of the most important sales of iron ever consummated in the state were, to a great degree, brought about by the fact that the laboratory could be depended upon not only for the inspection of the product, but also and particularly for the inspection of the stock.

Uniformly good iron can not be made at a uniformly low cost with irregular stock, and variations in the cost of the iron are to a considerable extent due to variations in the composition of the raw materials. Pay close attention to what goes into the furnace and the tapping hole will take care of itself. This is the key note of the

GENERAL DISCUSSION.

entire harmony. But there has not yet been a very full orchestra in this State, partly because ore was plentiful and cheap and partly because it seemed to be more economical to fill the furnace with almost anything that might be to hand and trust Providence to look after the cast-house.

There is nothing in the nature of the ores used that forbids their sale on analysis, and as this system is already applied to nearly all the flux used, and to a not inconsiderable quantity of coke and ore, the extension of it would not appear to offer insurmountable difficulties. The greater part of the cost of making iron is borne by raw materials. The quality of these materials, therefore, and their regularity of composition are of vital importance. As respects composition, there is a point beyond which it is not possible to make iron profitably, no matter what the price of the materials may be. How low this point may be will depend, *ceteris paribus*, upon the difference between the cost of the iron and its selling price. When this difference is considerable, as was the case in this State ten or fifteen years ago, iron may be made at a profit from very inferior materials. But when the margin of profit is narrow, as has been the case of late years, the use of inferior materials becomes impossible. With increasing competition and a narrowing selvage of profits, the necessity for using better and better ore becomes more and more pressing. To keep the furnaces in blast and avert disaster from the district, it may happen that the price of ore will fall below the figures at which it can be mined profitably, unless the operations be conducted on a very large scale and long time contracts can be made, assuring a steady output for a number of years. Under such conditions some concessions may be made by the furnace men in respect to quality, but at the same time they would be warranted in holding out for uniformity of composition. One would be inclined to consider the uniformity of composition as more important than the quality, provided always that this would not entail too much handling of stock per ton of iron made. When ore is sold for stock-house delivery at a fraction over a cent per unit of iron, it would seem that no further reduction in price could be expected.

Under all circumstances, except such as embody the sale of the ore at so much per unit of iron, there will be complaint by the furnace man that the ore is not as good as it might be, and it will be met by the miner with the assertion that it is as good as it can be at the price paid for it. This may, indeed, be true, but at the same time it is not to be hastily concluded that for more money the miner is willing to guarantee better ore. For the most part his endeavor is to get the largest possible returns from the smallest possible outlay, a resolution in the highest degree laudable but apt, at times, to cause more or less friction as to shipments. To him a ton of ore is a ton of ore. It weighs 2240 pounds, and whether it contains fifty per cent. of iron or forty-five he receives the same pay. But to the furnace man, who has to consider the amount of iron he can get from that ton and the ease with which he can do it, the question is of another kind.

There is a side of the matter not yet touched upon, but which can not be neglected. If the higher grade ore only is mined, the exhaustion of the deposit is certainly set forward. It rarely happens that all of a deposit is high grade ore, and if only the best is in demand one has to cut his cloth to suit the pattern. The miner may have incurred large expense in opening the mine and in equipping it with proper machinery under the expectation that his output would be profitable to him. If he is restricted to a certain portion of the ore and this be below the amount required to yield a profit on the investment, he would be subjected to hardships not tolerable under ordinary conditions. He is quite willing to encourage the belief that it is cheaper to use a large amount of low priced, low grade ore than to pay more for better ore of which not so much is used. In the minds of some whose opinions should be worthy of consideration the value of a fifty per cent. ore is proportional to the value of a forty-five per cent. ore, and they argue that as the lower grade material can be bought for fifty cents per ton, or 1.11 cents per unit of iron, the better grade material is worth proportionally more, or 55.5 cents per ton. They forget that the value of an ore increases very rapidly as one nears the fifty per cent. mark. As a matter of fact, if a forty-five per cent. ore is worth fifty cents, a fifty per cent. ore is worth 83 cents, that is, it will cost as much to make a ton of iron from the one at 50 cents as from the other at 83 cents. Above fifty per cent. the difference becomes even more striking.

Attempts at improving the quality of the ores used in the State have been confined so far almost entirely to the brown ores, although it is possible to better the soft red ores to a very considerable extent also. A description of the methods in use will appear under each kind of ore, so that it is merely necessary here to direct attention to the matter in a general way.

The ore that most readily lends itself to processes of beneficiation, without any very heavy expense, is the limonite or brown ore. Occurring, as it does, as more or less isolated masses imbedded in clay, it was comparatively easy to devise machinery that would treat the entire mass of stuff, removing the clay by suspension in water and passing the cleaned ore over screens of appropriate sizes. In this manner the clay, unless it was of a very plastic nature, was removed from the ore, the wash water being collected in settling dams and again used, after the clay had been deposited. The process was crude at first and the ore was insufficiently cleaned, but of late years it has been much improved and can now be depended on to furnish fairly good ore from even the more tenacious clays.

At some establishments it has been customary to improve the brown ores still further by calcining the washed ore in open piles with wood or charcoal "breeze" as fuel, and, later, in gas fired kilns. In this manner the ordinary water is completely removed, and the combined water, which does not go off under a full red head, to an extent depending on the temperature and the duration of the firing. Washed brown ore carrying 44 per cent. of iron can be greatly improved by calcining, the iron in the calcined ore being as high as 51 per cent. over a period of several months.

While it is now customary to wash nearly all the brown ore used in the State, but little calcining is done. The reasons for this practice will appear under the discussion of the brown ores, and it will be shown that unless the deposit is known to be large or the demands upon it not very exacting as to quantity, the erection of calcining kilns could not be expected to yield much return on the investment.

For improving the soft red ores several plans have been proposed, but none of them have worked their way into actual use on a large scale, although at least one of them may now be said to have passed the experimental stage. It was proposed to wash the lower grade soft red ores in such a manner as to remove the more ferruginous material from the more sandy portion and to recover the ore in settling dams. Some experiments were very successful as regards the possibility of concentrating the ore, but the large amount of water required at points where it was expensive to get and the impracticability of handling large quantities of damp ore that would certainly fall into the finest powder as soon as it was charged into the furnace have caused the investigation to be postponed.

During the last two or three years extensive experiments have been made with the hope of concentrating these ores magnetically. Two plans have been proposed. First, to render the ore magnetic by raising it to a full red heat in a properly constructed kiln and then passing a reducing gas over it so as to convert the ferric oxide into the magnetic oxide. Subsequent crushing and sizing would bring the ore into a condition in which it could be treated over a magnetic separator, the sand, &c. being removed by centrifugal action. A great deal of work has been done in this direction and the possibilities of the process are extremely encouraging.

The other plan for magnetic concentration of these low grade soft ores is to dry them thoroughly, crush and size and pass over a magnetic belt which will pick up the more ferruginous portions and allow the more sandy portions to fall away into suitable receptacles. Some work has been done along this line and the results are promising. It will be some time before definite information can be given to the public.

On the whole, therefore, it may be said that in actual practice the only ores subjected to a process of beneficiation on a large scale are the brown ores. Practically all of the pig iron made in Alabama is obtained from native ores. In this respect the situation is quite the reverse of that found in Ohio, which with a pig iron production of 1,463,789 tons in 1895 probably did not derive more than 3% of it from native ore. The only ores brought into Alabama for any purpose are some brown ore from Georgia, a little "spathite" ore from Tennessee, and Lake ore for use as "fix" in the rolling mills.

The production and value of the ore mined in the State, so far as can now be ascertained, are given in the following table, compiled from the reports of Mr. John Birkinbine to the United States Geological Survey, Division of Mineral Resources, from the census returns and from independent sources. For convenience of comparison the rank of the State as a producer of iron ore and the amounts and value of ore mined in the entire country are also given, for the same period.

IRON MAKING IN ALABAMA; THE ORES;
GENERAL DISCUSSION.

TABLE I.
PRODUCTION AND VALUE OF IRON ORES IN ALABAMA AND THE UNITED STATES.

	ALABAMA.				UNITED STATES.			Rank of Alabama.
	Tons.	Value.		Per cent. of Production.	Tons.	Value.		
		Per Ton.	Total.			Per Ton.	Total.	
1850	1,838	$3.68	$6,770	0.12	1,579,318	$4.23	$6,981,679	19
1860	3,720	5.31	19,765	0.15	2,401,485	5.31	12,757,848	15
1870	11,350	2.66	30,175	0.21	5,302,952	5.63	29,843,420	16
1880	171,139	1.18	201,865	2.3	7,497,509	3.09	28,156,955	7
1881	220,000	1.30	286,000	2.4	9,094,369	2.97	27,000,000	
1882	250,000	1.20	300,000	2.8	9,000,000	3.60	32,400,000	
1883	385,000	1.20	462,000	4.6	8,240,594	3.00	24,750,000	
1884	420,000	1.00	420,000	5.1	8,200,000	2.75	22,550,000	
1885	505,000	1.00	505,000	6.6	7,600,000	2.50	19,000,000	
1886	650,000	0.96	624,000	6.5	10,000,000	2.80	28,000,000	
1887	675,000	0.96	648,000	6.0	11,300,000	3.00	33,900,000	
1888	1,000,000	0.96	960,000	8.3	12,060,000	2.40	28,944,000	
1889	1,570,000	0.96	1,507,200	10.9	14,518,041	2.30	33,351,978	2
1890	1,897,815	1.00	1,897,815	11.8	16,036,043	2.20	35,279,394	2
1891	1,986,830	1.00	1,986,830	13.6	14,591,178	2.10	30,641,473	2
1892	2,312,071	1.06	2,442,575	14.2	16,296,666	2.04	33,204,896	2
1893	1,742,410	1.86	1,490,259	15.0	11,587,629	1.66	19,265,973	2
1894	1,493,086	0.83	1,240,895	12.6	11,879,679	1.14	13,577,325	3
1895	2,199,390							

For a number of years Michigan has held the first place as a producer of iron ore, Minnesota coming up from the 6th place in 1890 to the second place in 1894 and 1895. To show the disparity between the States ranking first, second and third since 1889, we need only glance at the following list:

	Michigan.	Alabama.	Pennsylvania.	Minnesota.
		Tons of 2,240 lbs.		
1889	5,856,169	1,570,000	1,560;234	
1890	7,141,656	1,897,815	1,361,622	
1891	6,127,001	1,986,830	1,272,928	
1892	7,543,544	2,312,071		1,255,465
1893	4,668,324	1,742,410		1,499,927
1894	4,419,074	1.493,086		2,968,463

Alabama held the second place from 1889 till 1894, when she was surpassed by Minnesota, and Pennsylvania the third place until 1892 when Minnesota came up to the second place. It is not likely that the relative positions will be changed for some years. The immensity of the Mesabe ore deposits and the cheapness with which they are mined will, perhaps, keep Minnesota in the second place for the next decade, if indeed she does not push Michigan for first place within that time. Michigan does not produce much pig iron, the output being 91,222 tons in 1895. Minnesota made no iron in 1894, nor in 1995. The difficulty of procuring good coke at that distance from the coal fields has hitherto prevented these States from converting their ore into iron, and the tendency seems to be more and more to reduce the cost of these ores to Illinois, Ohio and Pennsylvania furnaces. But it is a wise man who prophesies concerning the iron trade in this day of rapid industrial changes. It would appear, however, that Alabama will have to face competition from furnaces much nearer than Michigan and Minnesota. It is just here that questions of transportation play the really vital part. So long as the rich Lake ores can be hauled to Ohio and Pennsylvania furnaces and converted into pig iron which can be sold profitably for half a cent per pound, the situation in Alabama will be one in which the cost of transporting the iron to market after it is made is the main question. With the Northern and Eastern furnaces the great question is the cost of gathering the raw materials into the stockhouse. In Alabama the great question is the cost of marketing the pig iron. With better ore, better coke and better furnace practice it may be possible even in Alabama to reduce the cost of making iron, but the transportation companies will control the situation then as they do now, unless a closer union can be effected between the two interests. We can not hope to avail ourselves of water transportation on a larger scale, as is done in the case of the Lake ores to Illinois, Ohio and Pennsylvania ports. In providing cheap ore, cheap coke and good flux within short distances of each other, nature seems to have thought that she had done enough for Alabama, and failed to provide water-ways for conveying the product to market; an oversight much to be deplored, indeed, but to be accepted with becoming fortitude. -

According to the Cleveland Iron Trade Review, Cleveland, Ohio, the Lake shipments of iron ore in 1892, were 8,545,313 tons; in 1893, 5,836,749 tons; in 1894, 7,621,620 tons; and in 1895, 10,234,910 tons. These figures mean that considerably more than half of the total amount of iron ore mined in the United States is transported by water to the vicinity of the furnaces using it. Were it not for this fact the enormous development that has been reached in the Lake regions, with respect to the mining of iron ore, could not have been attained within so short a time, if at all.

In order to exhibit the relation that Alabama sustains to the other iron ore producing States, both in respect to the amount mined and the value, the following table, taken from the report of Mr. John Birkinbine in 1895, to the U. S. Geol. Survey, Division of Mineral Resources, is appended.

TABLE II.

Total Valuation and Average Value Per Ton of Iron Ore Produced in the United States in 1889, 1892, 1893, 1894.

	1889		1892		1893		1894	
	Valuation.		Valuation.		Valuation.		Valuation.	
	Total.	Per Ton.	Total.	Per Ton.	Total.	Per Ton.	Total.	Per Ton.
Alabama	$ 1,511,611	$ 0.96	$ 2,442,575	$ 1.06	$ 1,490,259	$ 0.86	$ 1,240,895	$ 0.63
Colorado	487,433	4.47	587,903	4.15	514,312	3.00	676,141	2.70
Connecticut and Massachusetts	a 265,901	3.01	249,198	3.27	122,475	3.01	71,191	2.35
Georgia and North Carolina	334,025	1.29	262,517	1.25	203,682	1.09	166,228	0.95
Kentucky	135,559	1.75	63,172	1.25	47,746	1.30	54,379	1.28
Maryland	f 68,240	2.32	88,691	2.21	25,585	1.85	17,809	2.25
Michigan	15,800,521	2.70	16,587,521	2.20	8,611,192	1.84	5,844,995	1.32
Minnesota	2,478,041	2.87	3,090,942	2.46	2,321,204	1.55	2,165,802	0.73
Missouri	561,041	2.11	237,827	2.01	160,532	2.07	105,235	1.28
Montana, N. Mexico and Utah	b 269,164	3.12	c 97,121	2.16	d 103,545	2.67	e 67,538	1.52
New Jersey	1,341,543	3.23	1,388,875	2.98	909,458	2.55	568,056	2.05
New York	3,100,216	2.49	2,379,267	2.67	1,222,934	2.29	396,456	1.63
Ohio	532,725	2.09	148,288	2.01	104,897	1.54	65,702	1.12
Pennsylvania	3,063,534	1.96	2,197,028	2.03	1,374,313	1.97	643,459	1.21
Tennessee	606,476	1.28	505,359	1.24	392,771	1.05	288,005	0.98
Texas	19,750	1.52	20,890	0.91	25,997	1.01	11,521	0.75
Virginia and West Virginia	935,290	1.83	1,428,801	1.91	1,050,977	1.70	873,305	1.45
Wisconsin	1,840,908	2.20	1,428,921	1.81	584,094	1.33	320,518	0.92
Total	$33,351,978	$ 2.30	$33,204,896	$ 2.04	$19,265,973	$ 1.66	$13,577,325	$ 1.14

a. Including Maine.   b. Including Oregon, Washington, and Idaho.   c. Including Oregon.   d. Including Oregon

CHAPTER II.

THE HEMATITE ORES.

In the discussion of the hematite ores we shall have to exclude the brown hematites as they properly belong to the limonites, although often mis-called by the former name. The limonites are locally termed "brown ores" and the output is about 25 per cent. of the total ore production of the State. They will be discussed under their proper heading.

The hematite ores are, for convenience, classed under two heads:

First, the soft red ores, carrying but little lime and

Second, the "hard red" ores carrying from 12 to 20 per cent. of lime and in many cases self-fluxing, that is, they carry enough lime to flux the silica contained in them.

In order that a clear understanding of the matter may be had at the outset the following brief description of the geological and topographical features of the deposit of hematite ores so largely used in the State is given here.

They belong to the Clinton formation of the Silurian, which extends with some breaks, from the middle portion of Alabama to the northern part of Maine. They are overlaid by chert, sandstones and clays, the overburden at places reaching a depth of forty and fifty feet. The seams now worked vary in thickness from 3 to 25 feet, run in a north-east direction and dip towards the south-east at angles varying from 15 to 22 degrees, the dip increasing as one goes towards the south-west. For the most part they occupy the crests of the hills, the outcrop forming a striking and persistent feature of the landscape for several miles in the vicinity of Birmingham.

The Soft Ores.

As a rule, to which, however, there are some important exceptions, the outcrop is "soft red," a term of comparative significance as the ore is quite firm and has to be won, by regular blasting operations. It is soft as compared with the limey or "hard red" ore. The soft ore may extend from the outcrop for a distance of 300 feet on the dip, depending on the thickness and imperviousness of the cover although the hard ore comes to the surface at more than one place.

In winning the soft red, the overburden is removed and the ore mined, at day, by benches. Under cover the ore becomes limey and hard and is mined from inclines on the dip by drifts and slopes.

The soft ore is the hard ore with the lime removed by atmospheric influences and is richer in iron the poorer it is in lime. When the overburden is stripped off there is found a seam of ore quite soft and seemingly disintegrated, of a deep red or purple color, the so-called "gouge." It may be only a few inches thick but often runs to 24 and even 36 inches, and comprises generally the best part of the ore. Underneath this begins the more solid ore diminishing in content of iron according to the vertical depth. The best quality of "gouge" will carry 52 per cent. of iron while ten feet below its line of demarkation the iron falls to about 46 per cent. Between the "gouge" and the ore proper there is often a thin seam of yellowish clay, which, however, is by no means constant in strike. In the more solid ore, beneath the "gouge" there are seams of the same clay, sometimes as much as two inches thick but for the most part not above half an inch thick. In the early days of iron making in the Birmingham district it was the custom to mine from 12 to 20 feet of the soft ore and to send the whole material to the furnace. Of late years, however, the mining has been restricted to ten feet, including the "gouge," as it was found that below this depth the ore rapidly became siliceous and unfit for use. Taking the content of metallic iron in the "gouge" at fifty per cent. as mined, the loss in iron according to vertical depth, is about one-half of one per cent. per foot. This would bring the iron in the first ten feet of the seam to forty-five per cent. and in the next ten feet to forty per cent. A large number of analyses extending over several years show that when the mining is limited to the ten foot mark the iron content is a little over 47 per cent. in the ore as mined, i. e. with seven per cent. of water, and including the "gouge." The rapid increase of the silica in the ore below the ten foot mark is shown by the fact that to get even 47 per cent. of iron in the upper ten feet from one-fifth to one-third of it must be composed of the "gouge," with its 50 per cent. of iron.

In Vol. XV. 10th U. S. Census, Mr. A. A. Blair gives some very detailed analyses of the soft red ore used in the Birmingham district. An average of those quoted is herewith given:

	Dry basis %
Silica	13.66
Sulphur	0.11
Phosphorus	0.43
Alumina	6.13
Lime	1.26
Magnesia	0.37
Manganese protoxide	0.30
Iron protoxide	0.32
Iron peroxide	75.05
Carbonic acid	0.08
Carbon in carbonaceous matter	0.03
Water of composition	1.62
	99.36

Metallic iron, 52.87 per cent.
Specific gravity 4.

This average shows a greater amount of alumina, and metallic iron, and much less silica than is usually the case with this class of ore.

An average analysis of stock house samples shows:

		Dry.
Iron	47.24	50.80
Silica	17.20	18.50
Alumina	3.35	3.60
Lime	1.12	1.20
Water	7.00	

For practical purposes it is not necessary to go so fully into detail and it is customary to determine merely the insoluble matter and the iron. With a few ores of this class, which carry unusual amounts of alumina this ingredient is also determined. But for every day practice and with slags of 33 to 36 per cent. silica the alumina is considered as silica and reported with it as "insoluble." It is a fortunate circumstance that the soft red ores, when finely ground, yield their iron to acid solution without fusion, the insoluble residue being of a creamy white appearance and carrying seldom more than 0.20 per cent. of iron. For blast furnace purposes and where the ore is not sold on the unit system the easy solubility of the ore is a point of great importance, especially when many analyses must be made within a short time. About one-half of the alumina present goes into solution with the iron but may be neglected under the conditions that obtain in the district with respect to the variation in the composition of the cinder. In calculating furnace burdens the error arising from neglecting the alumina and reckoning it as silica is comparatively slight, as the ratio between the silica and the alumina is as 1 to 0.87.

The insoluble matter in most of the soft red ore as used in the state is 23 per cent., and the iron 46 per cent., with water at 7 per cent. The ordinary ratio between the metallic iron and the insoluble matter varies from 1 to 1.50 to 1:2. To illustrate:

Water 7%.

Iron.	Insoluble Matter.	
40	35.00	
41	33.00	
42	31.00	For each 1 per cent. increase in the Iron the Insoluble Matter falls 2 per cent.
43	29.00	
44	27.00	
45	25.00	
46	23.00	
47	22.00	
48	20.50	
49	19.00	For each 1 per cent. increase in the Iron the Insoluble Matter falls 1.50 per cent.
50	17.50	
51	16.00	
52	14.50	
53	13.00	
54	11.50	

It is not necessary to carry the list further, as the supply of fifty-four per cent. soft red ore is limited. It is not claimed that this ratio is absolutely correct, but a large number of analyses substantiate its reliability for ordinary purposes. The ratio from 40 iron through 46 iron is as 1:2. Beginning with iron 47 and insoluble 22, the ratio appears to be nearer 1:1.50 than 1:2, for with iron 48 the insoluble matter is about 20.50. It may, therefore, be said with a fair degree of accuracy that a soft red ore carrying 40 per cent. of iron may be expected to contain 35 per cent., one with 45 per cent. of iron 25 per cent., and one with 50 per cent. of iron 17.50 per cent. of insoluble matter. There are, of course, exceptions to this rule and it does sometimes occur that an ore with 46 per cent. of iron will be found to carry 22 per cent. and one with 48 per cent. of iron will have 21 or 22 per cent. of insoluble matter.

But on the whole the fact remains that an ore with 45 per cent. of iron will carry 25 per cent. of insoluble, and one with 50 per cent. of iron from 17 to 18 per cent., and the list may be used as an approximation to the truth.

In texture the soft red ore is a mass of minute siliceous pebbles held in a ferruginous cement. The pebbles are seldom larger than a No. 4 shot, and are frequently much smaller. They are all more or less rounded and stained reddish-brown. The cementing material is softer than the pebbles, and on sizing even a very lean ore the material passing a screen of fifty meshes per linear inch is much richer in iron than the material remaining on a 10 or a 20 mesh screen. A soft red ore of 40 per cent. iron, on being ground to pass a ten mesh screen, will yield through a fifty mesh 53 per cent. of iron.

So far as concerns their physical structure, this is one of the points of differentiation between the soft red and the so-called brown ores, for these, on being sized, show a steady loss of iron the finer the screen. The fact of increasing richness in iron the finer the screen renders the concentration of the low grade soft red ores much simpler than would otherwise be the case, as the "fines" can be briquetted without further treatment, and the troublesome question of handling them becomes comparatively easy. The rounded form of the more siliceous pebbles also occasions less wear on the shutes, screens and conveyors, a point of no little moment in concentrating works.

The better grades of the soft red ore do not occur at every point on Red Mountain, nor is it possible to mine even ten feet profitably everywhere along the ridge. It is frequently the case that the inferior ore sets in, as the saying is, "at the grass roots," and even the richer "gouge" is sometimes absent. Mining operations can not be undertaken without careful prospecting and many analyses, for the difference between a fairly good ore and one that is not passable is often so slight as to deceive even the most experienced man who grades merely by the eye. After having become accustomed to a particular kind of ore, one may judge of its quality by the appearance with a reasonable degree of accuracy. While for the most part the soft ores are of the same general texture and color, it not infrequently happens that serious mistakes may be made unless the services of a chemist are called into requisition. When freshly mined the ore is of a deep red color, inclining to purplish red in the richer portions, but on drying there is assumed something of a brownish tint. For ordinary stockhouse delivery the ore contains on the average 7 per cent. of hygroscopic water, which, owing to the coarse-grained nature, soon dries out under cover.

In the early days of iron making in the Birmingham district, before the real value of the limey or hard ores was generally accepted, the furnace burden was composed almost entirely of the soft ores. Of late years, however, the tendency is decidedly towards a greater and greater proportion of the limey ore, the proportion rising at times to above 90 per cent. of the ore burden. It is still to some extent a mooted question as to the relative reducibility of the two ores, but a careful investigation of the subject would, we think, show that in this respect the limey ore has the advantage. When the soft ore descends into the zone of reduction in the furnace, it does so without losing its firmness of texture. Even after it has become red hot, or white hot, it maintains its shape, except as this may be changed by friction during the descent. The reducing gases act upon it in the lump, and if the lumps be of considerable size the reduction to metallic iron may be delayed and the ore may appear before the tuyeres.

The case is quite otherwise with the limey ore. The

lime is present as carbonate, (except such as may be combined with the prosphorus as phosphate of lime, an amount rarely exceeding 0.50%,) and when this reaches a point in the furnace at which its carbonic acid begins to come off, the ore begins to fall to pieces. The friction of the other materials aids this tendency quite as much as, and perhaps, more than in the case of the soft ore. The reducing gases can and do have a greater ore surface to work on and the result is that for a given weight of coke and a given composition of the gas there is greater reducing action. The soft ore is more fusible than the lime ore, but this does not necessarily mean that it is more easily penetrated by the reducing gases within the furnace. On the contrary a fused crust on the outside of a piece of soft ore interposes considerable opposition to the passage of the gases, and as this crust becomes thicker and thicker the gases penetrate with more and more difficulty. In the case of the lime ore as soon as it begins to part with its carbonic acid it begins to disintegrate, and this very fact of disintegration enables it to receive to better advantage the reducing power of the gases.

In comparing the two ores another circumstance must not be lost sight of, and that is the intimate commingling of the ore and the lime that is to flux it. This is a distinguishing characteristic of the lime ores. It would be impracticable to effect by artificial means such an intimate mixture of ore and lime as Nature has already provided in these ores. This circumstance is of the greatest importance in any discussion of the relative value of the soft and the lime ores, for while these latter require a higher heat for fusion they are not therefore to be considered less easily reducible.

The reducibility of an ore depends far more upon its permeability or porosity than upon its fusing point. For the most part the loss of energy in a furnace is chargeable to lack of reducing power rather than to lack of fusing power.

The tendency now is more and more towards the use of the limey ores; for the enormous demand that has been made on the better quality of the soft ore within the immediate vicinity of Birmingham is beginning to make itself felt,

Three courses of action may be open: First, the increasing proportion of limey ore in the burden may induce the furnacemen to look towards the use of eighty or ninety per cent. of it, the difference being made up with soft and brown ore. Second, other sources of soft ore may be utilized. Third, the lower grades of the soft ore, now remaining in the ground, may be concentrated and made to take the place of the ore that has been removed. It is not thought that the proportion of brown ore used will be materially increased.

Under the existing conditions it would appear advisable to begin at once to increase the proportion of limey ore used, so as to establish on the basis of wider experience the economic relation that this burden would sustain to former practice, or to push the work of concentrating the lower grades of soft ore to some definite result.

The experiments on concentrating soft ore, to which some allusion has already been made, showed the possibility of taking an ore of 40% iron and 35 % silica and bringing the ore to 57% and the silica to 15%, on the average. In this process two tons of raw ore were required to make one ton of concentrates. Up to this time 200 tons of concentrates have been made and the experiments are still in progress. Results so far reached indicate that under proper conditions the cost of one ton of concentrates of the above given composition would approximate $1.00, putting the raw ore at 30 cents per ton at the works.

The process was described by the writer in a paper read before the American Institute of Mining Engineers at Atlanta, Ga. October, 1895. Since that time much additional information has been gathered concerning it. In brief the process is to magnetize this ore by heating it to redness in a suitably constructed kiln and passing producer gas over it. In this manner the ferrice oxide is converted into magnetic oxide and by crushing the ore, screening and treating over a magnetic separator the more siliceous material is eliminated. The richer portion of the ore thus improved carries nearly 60% of iron with 11% of silica. The ore treated in this manner can be sent to the furnace direct or made into briquettes or eggettes with tar or other binding material. If sent direct to the furnace it would be of such fineness as to pass a screen of ten meshes per linear inch, about 25% by weight passing a forty mesh screen. In respect to its fineness, therefore, it would be coarser than a great deal of the Mesabe ore now consumed in Ohio and Pennsylvania furnaces, while in content of iron it would, on the average, carry about five per cent. less than this Mesabe ore. The soft ore as now used carries about 47% of iron, and the Mesabe ore from 62% to 64%. Several hundred thousand tons of low grade soft ore are now uncovered in the Birmingham district, forming that portion of the big seam from which the upper ten feet have been removed and carrying about 39% of iron. It is easily and cheaply mined and lends itself very well to concentration. This is the ore which should take the place of the soft ore now being mined and it is much to be hoped that the work already taken in hand in respect to it will be carried to completion. The importance of the matter is assuredly great enough to warrant the expense required. There is no doubt at all of the possibility of concentrating this ore and no doubt of the value of the product to the furnace. The successful prosecution of this business would bring into use very large amounts of ore that can not be used unless concentrated and would prolong the life of the soft ore for many years.

The possibility of concentrating this ore magnetically, without previous magnetization, is now under consideration, and the results reached are of the most encouraging character.

The Limey, or so-called Hard Ore.

The ore sets in sometimes at the outcrop but much more frequently it is found only under cover and is the continuation of the soft ore in the direction of the dip. For distances varying from nothing to 300 feet on the dip the ore is soft, then the hard ore begins and continues to depths not yet ascertained but certainly very considerable. In other words, as has been already stated, the hard ore which originally appeared at the surface has been deprived of its carbonic acid by atmospheric influences and converted into soft ore along the dip to varying depths the lime having been removed by bleaching. Relatively the same differences that are to be observed in the soft ore from various places are also found in the hard ores. There are points along the mountain where the minable seam of soft ore is better than at others, and there are places where the hard ore is better than at others.

On a vertical section of the soft ore the content in iron decreases downward, the rate being about one-half of one per cent. per foot. The rule holds good for the hard ore on a vertical section. The mining on the big seam of soft ore is now confined for the most part to the upper ten feet, the mining on the hard ore is also the same, and below the ten foot mark the hard ore also becomes too siliceous for economic use. The hard ore derives its value from two circumstances, first there is a great deal more of it than of the soft ore, because it extends to very considerable depths, and second because of the intimate admixture of carbonate of lime with the ferruginous material. The best hard ore carries more lime than is required to flux its silica while in the ordinary grades the ratio of one of silica to one of lime is generally conserved. When this is the case the ore is termed "self fluxing" and in burdening a furnace exclusively with hard ore of this type it is not necessary to add limestone. When the burden is composed of hard and soft ore, or of hard and brown, or of hard, soft and brown the amount of limestone to be added is calculated from the silica of the ore other than hard, the silica of the fuel and of the stone itself. The increase in the use of hard ore would tend to diminish the consumption of limestone by an amount represented by the limestone in the ore and if a strictly self-fluxing ore were used the consumption of limestone would be greatly diminished. There is a kind of hard ore, termed semi-hard, which contains from one-third to one-half of the lime in a typical hard ore, but of this sort very little is used and it is not mined regularly.

Within the last two years the use of crushed hard ore has become quite common in the Birmingham district. The soft ore does not lend itself readily to crushing unless thoroughly dry. With the amount of water it usually contains it becomes somewhat like clay in the crusher, i. e. more or less gummy, and the machine soon become choked.

A general average of the hard ore used shows:

	PER CENT.
Water	0.50
Metallic Iron	37.00
Silica	13.44
Lime	16.20
Alumina	3.18

Phosphorus.................................... 0.37
Sulphur....................................... 0.07
Carbonic acid.................................12.24

Adding the alumina and the silica together we have for silica x alumina 16.62%, the lime is 16.20%, and the ore may be termed self-fluxing. It can not be said that all of the hard ore used is self-fluxing, as some of it contains 5% more of lime than of silica x alumina. Taking a general average, however, of analyses of all kinds of hard ore extending over several years this ore carries enough lime to flux the silica x alumina. It may be urged that aluminous soft ore needs silica as a flux for the alumina, and this is indeed true. But we have to flux the silicate of alumina with lime, and it is merely a question as to whether all the bases of the burden shall be calculated as lime, and all the acids as silica or whether we shall regard the silica x alumina as requiring so much lime. In either case the type of slag to be made has to be considered, and for any one type the two calculations lead to the same result so far as concerns the consumption of limestone per ton of iron.

The question has been raised as to whether the hard ore, on the dip, may not gradually lose its content of iron and become a more and more ferruginous limestone until finally the iron will not exceed 20 or 25%. The matter is one of scientific rather than practical moment, and some information has been collected. Taking the iron in the soft ore at 47% at the outcrop, and in the hard ore at 37% 100 feet on the dip the rate of decrease for the iron would be one per cent. per hundred feet. This rate seems to be maintained at some localities, but at others it varies so that no rule can be given. This comparison is between the soft and the hard ore. When the hard ore begins it maintains a fairly uniform composition on planes extending in the direction of the dip.

As to the minimum amount of iron that a hard ore can carry and still be considered an ore, opinions may differ. But if the iron in the hard ore should fall to 25%, the lime increasing in the same proportion, it is not likely that it could be used. The silica and alumina appear to remain somewhat stationary, so that the question would be whether or no material carrying 25% of iron, from 16 to 20% of silica, and from 24 to 28% of lime can be profitably used. It will be many years, however, before this question will arise, and it is not necessary to discuss it now. It is bound up with geological and topographical considerations which are still in abeyance. Some work has been done in the direction of improving the hard ore by calcining it in a gas-fired kiln. It is possible in this way to remove the carbonic acid entirely. Taking the average analysis of hard ore as given, viz:

Iron...37.00
Silica..13.44
Alumina...................................... 3.18
Lime...16.20

and considering all of the lime as carbonate except 0.50% as phosphate, the carbonic acid would be 12.24. Removing this the above analysis would show:

Iron...42.15
Silica..15.31
Alumina...................................... 3.62
Lime...18.46

The ore would, of course, still be self-fluxing, and the question would be whether the removal of the carbonic acid outside of the furnace, with the consequent transformation of the carbonate of lime into caustic lime, would benefit the ore more than it would cost.

Without entering upon any lengthy discussion, as the matter has not yet passed the experimental stage, we may regard the question briefly, from a physical and a chemical standpoint.

Physically the ore would become more porous as the expulsion of the carbonic acid would, to a great extent, destroy its compactness. It would lose in weight, but this would be more than counter-balanced by the gain in the per centage of iron. Its increased porosity would allow easier penetration for the reducing gases of the furnace. Against this may be placed its increased friability, and the consequent production of a greater quantity of fine material in the furnace. Chemically, we should have to consider the effect upon the combustible gases of the introduction of caustic lime instead of carbonate of lime.

The carbonic acid has to be removed and the question narrows down to a single consideration, viz. Is there any advantage in removing it outside of the furnace? The heat within the furnace removes it quite as effectively as the heat of a kiln, but then we would have to weigh the effect of large volumes of hot carbonic acid on the coke, with solution of carbon, &c. Cokes differ markedly in this respect, and each one has to be examined in and for itself. If the calcined ore is charged direct it would carry a considerable amount of heat into the upper part of the furnace and it would be more difficult to maintain a cool top. This, however, need hardly be considered, as the additional temperature, due to charging hot material, would be derived, not from reactions within the furnace, but from extraneous sources. A cool top under ordinary conditions means that the heat within the furnace is used in melting the stock, and is not escaping in the gases. But if a hot top is due to extraneous heat, such, for instance, as hot material charged, there would be no injurious effect upon the zone of fusion. It might be advantageous to have a hot top if the heat was not derived from the reactions within the furnace, as the gases to be consumed under the boilers and in the stoves would arrive at the burners at a higher temperature. Aside from such considerations, however, it seems advisable to use the calcined ore direct. Where it is stocked, or allowed to remain even for twenty-four hours in the air, it rapidly takes up water and becomes pasty. When the slacking of the caustic lime is completed the material appears dry but in reality contains not only water of hydration but carbonic acid also. When the water of hydration is expelled the lime becomes pulverulent and dusty, blows about in every breeze and is troublesome to both bottom and top fillers. It can be dampened with water from a hose-pipe, of course, but in that case the mass becomes pasty, and the stockhouse uncomfortable. If the ore is not used direct, (the kiln being in immediate proximity to the furnace), the advantages to be obtained from calcining begin to disappear at once, and continue to become less and less the longer the interval between calcination and charging.

THE LIMONITE, OR SO-CALLED BROWN ORES.

As a rule these ores constitute the best material for iron making in the State. Practically all of the charcoal iron is produced from this class of ore, and although there has been of late years a marked decrease in the output of charcoal iron, following the general tendency throughout the country at large, the total amount made from 1872 to the close of 1895, was 1,191,145 tons.

The yearly amount of brown ore mined is about 25 per cent. of the total production of all kinds of ore.

The deposits do not occur in regular seams except as the gossan of underlying pyritiferous veins which furnish very little of the ore used, but as pockets in the clay. These pockets are of greater or less extent, some times going down to 75 or 100 feet, or even deeper.

They do not appear to follow any known rule of occurrence, and each deposit has to be judged by itself alone. It is a common saying that no one knows much about a brown ore bank beyond the length of his pick. To-day one may be in good ore, tomorrow there may be none in sight, and to know which way to turn one must know the particular deposit he is mining.

The ore is of two kinds, lump, and gravel. There is no rule as to the proportion in which each may be present, even in the same 'bank.' The lump ore is generally better than the ordinary gravel ore unless this latter is carefully washed from adhering clay. And yet it often happens that the presence of chert, or sandy inclusions, in the lump ore, as also the clay-filling of the interstices and small holes, makes the lump ore objectionable. The lumps vary in size from that of the fist to large masses of several tons weight.

The large lumps are broken by hand, if of unusual size by means of small charges of dynamite, and loaded on the car without further treatment. By far the greater amount of brown ore is comprised within the sizes of a pigeon's egg and a goose egg.

Excluding the large lumps the method of mining is briefly as follows: The bank is cut away in benches, the entire mass being taken down either by hand, or steam-shovel. The stuff is loaded on trams and conveyed to ordinary log-washers, single or double as the case may be, where it is subjected to thorough disintegration and stirring in large excess of running water. The clay &c. is removed by suspension in water, and is run into settling dams for the recovery of the water. The heavier particles of sand are screened out over $\frac{1}{4}$ inch screens revolving in a mild current of water, and the washed ore delivered over the screens into the railroad cars, and sent to the furnaces. Where the clay holding the gravel is friable, and does not 'ball' under the action of the washer, and where abundance of water can be secured, this method of preparing brown ore is fairly successful. There is great variation in the character of the clay, some of it being easily disintegrated and therefore yielding its ore readily, and some of it being extremely tenacious and putty-like. In this case there may be serious loss of the finer ore particles, the balls of clay picking them up, enwrapping them, and finally carrying them to the waste dump.

It is customary at some establishments to remove the clay balls by hand, boys being employed for the purpose. Jigging is resorted to but rarely, the results not warranting the additional expense.

A method of washing that has given good satisfaction is to discharge the trams from the 'bank' into a headbox in which play two powerful streams of water. The lower end of the box, which is of triangular shape and inclined about 30 degrees, opens into a long wooden trough lined with castings of iron fitted snugly at the bottom. This trough in turn discharges into the washer at the foot of the hill.

The advantages claimed are contact of the material with water under pressure, and the better separation of ore and clay from the tumbling motion down the trough. Even the tenacious clays may, in this manner, be made to yield their ore. But if the clay be extremely tenacious, as is sometimes the case, even this mode of treatment fails to disintegrate it. In fact it rather tends to increase the 'balling' by carrying the material down an incline. The friable and easily disintegrated clays, on the other hand, are speedily removed in this process, and the washer is called upon merely to complete what has been already pretty well done. No washing system can succeed without plenty of water, and unsparing use of it. If the best results are to be reached there must be no half-handed and mistaken economy in the consumption of water, and as a large part of the water used is recovered in settling dams the loss of water is chargeable mostly to evaporation and seepage. The first can not be prevented, but seepage can be controlled by properly constructed dams.

The amount of material moved per ton of ore obtained varies within wide limits. It may be 1 : 1, 4 : 1, or 10 : 1. Even the same bank shows very considerable differences in this respect, so that no rule can be given. It is a matter that can not be determined before hand, and is liable to change from day to day. Variations in the composition of the ore from the same bank, while observable, do not, as a rule, offer serious obstacles to successful mining. A given bank is apt to afford ore of the same general composition, and variations in the composition of stock-house samples are to be explained by insufficient treatment in the washer, due to lack of water or changes in the nature of the clay.

Brown ore mining is attractive because of the higher price paid for good brown ore, but should be entered upon only after the most thorough examination of all local conditions.

The average composition of the brown ore of the State, stock-house delivery, is as follows:

DRY BASIS.

Metallic Iron	51.00
Silica	9.00
Alumina	3.75
Lime	0.75
Phosphorus	0.40
Sulphur	0.10

The amount of water it contains varies according to circumstances. Thus, if the washer be placed at a short distance from the furnace the water, not having had time to drain out, is more than if the haul were longer. So also if the ore be not properly washed the clay retains water. Under a haul of 25 to 50 miles the ore, samples from the cars in the stock-house, contains on the average 7% of hygroscopic water. Following is an average analysis of a good quality of brown ore:

Hygroscopic water	7.00
Combined water	6.00
Metallic Iron	48.54
Silica	11.22
Alumina	3.61
Lime	0.84
Phosphorus	0.38
Sulphur	0.09

Selected brown ore may carry as much as 56% of iron, on a dry basis, and at one establishment the ordinary ore as charged carries 53%, after washing and calcining.

The sale of brown ore on analysis has become the custom in the Birmingham district for outside ores. The basis of sale is 50% of Iron, and 10% of insoluble matter, or silica, as the case may be. The price per ton is started, let us say, at $1.00, for ore carrying 50% of Iron, and 10% of Insoluble matter. Then for each one per cent. above 50% 5 cents per ton is added to the price. If the Insoluble matter at the same time decreases 1%, being 9% instead of 10%, 2½ cents per ton additional is added. An ore carrying 51% of iron and 9% of insoluble matter would be worth $1.075 per ton, and so on. If, on the contrary, the percentage of metallic iron should fall to 49%, 5 cents per ton would be taken off, and if at the same time the insoluble matter should rise to 11%, 2½ cents per ton more would be subtracted. Thus an ore carrying 49% of iron and 11% of insoluble matter would be worth $0.925 per ton. The starting price is not always the same. It may be $1.00, $1.05, $1.10 &c. according to circumstances, but the valuation of 5 cents per unit of iron, and 2½ cents per unite of insoluble matter is generally adopted. In this scheme no account is taken of hygroscopic or combined water, or of sulphur, phosphorus or alumina.

The basis of valuation is the amount of metallic iron and insoluble matter. The ore may contain 5%, or 10% of ordinary water, yet no account is taken of it. It would be much better if a deduction could be made for all water above a certain percentage, although the condition of the weather, as in the case of heavy rains while the ore was in transit, might prevent satisfactory agreements.

The water a brown ore may contain is a small matter compared with the clay it may, and too often does, contain. The ordinary water is easily enough evaporated in the upper part of the furnace, but the clay requires fuel and stone for its removal.

Well washed ore, free from clay, seldom holds more than 4% of water, and the increase in the amount of water follows closely upon the increase in the amount of clay.

There is a circumstance in connection with brown ore that merits attention, not only because of its contradistinction to the soft red ore but also and particularly because of its bearing upon its improvement, whether by simple screening or by some magnetic process. It has been stated that even the lower grades of soft ore on being dried and crushed yield more metallic iron in the material passing a 50 mesh screen than in the coarser stuff. In such ores there is a marked increase in the iron the finer the screen up to and including a 50 mesh. This is not true of the brown ore. The finer the screen, up to and including a 50 mesh, the poorer in iron is the material passing through.

Not only have laboratory experiments shown this but actual work on a large scale has substantiated the general truth of the proposition that on crushing brown ore, whether by machines, or by the attrition of ore on ore in a kiln the fine stuff carries less iron than the coarse stuff. Attention is drawn to this matter because of the custom at some kilns to draw the ore over screens into the furnace-buggies. There is considerable loss of material in this practice, and it is not to be recommended unless the ore carries an unusual amount of clay, which, of course, is removed over the screens. It may happen that as much as 10 per cent. by weight is lost, even over a $\frac{1}{2}$ inch screen. Some experiments were undertaken to establish the actual loss, and how much iron was present in the various sizes of ore from a kiln.

Several hundred pounds were taken, the samples being drawn over several days and put together, so as to represent the ore fairly. The results of the investigation were as follows:

	Iron.	Silica.
Raw ore	44.63	13.82
Calcined ore	50.20	15.10
Calcined ore—		
On ¼ inch screen (68 per ct.)	52.95	10.25
Through ¼ inch screen (32 per ct.)	49.30	15.90
On ⅛ inch screen (77 per ct.)	52.75	11.05
Through ⅛ inch screen (23 per ct.)	42.85	21.80

It can not, of course, be said that all brown ores act in this way, but the ore under examination fairly represented the second grade brown, and it is likely that other ores of the same class would give results comparable to these.

Screening over a ¼ inch screen gave 68 per cent. on the screen, with, say, 53 per cent. of iron, and 32 per cent. through the screen with 49.50 per cent. of iron. Screening over a ⅛ inch screen gave 77 per cent. on the screen with 52.75 per cent. of iron, and 23 per cent. through the screen with 42.85 per cent. of iron. Screening can not be recommended, except for clayey ore, and the clay should be removed in the washer. There is practically but little difference between the 'overs' on a ¼ inch and ⅛ inch screen in respect of iron, while there is a difference of 9 per cent. in weight in favor of the coarser screen. The loss of ore through either screen is too large for profitable work, except under unusual circumstances requiring the use of the best ore obtainable.

Reference has been made to the fact that for the most part the brown ores are washed but not calcined. In the production of charcoal iron it is the usual custom to wash and calcine, but as the consumption of brown ore in the charcoal furnaces from 1890, to and including 1895—probably did not exceed 7 per cent. of the total brown ore production during that period it can not be said that calcining is commonly practiced. When it is carried on two methods are used, the old fashioned open air pile fired by charcoal "breeze" and wood, and the new fashioned gas-fired kiln employing producer-gas as fuel. The former method needs no description. When properly managed it gives fair results, but can not be depended on to furnish uniformly calcined ore. Even with careful attention a part of the ore will not be calcined at all, part will be calcined properly, and part will be louped. The most curious mis-statements are sometimes made in reference to calcining brown ore, to say nothing of the idea, prevalent among some who ought to know better, that brown ore is termed limonite because it contains considerable quantities of lime.

In the hearing of the writer, the general manager of an iron company stated to a party of capitalists who were examing the property, that on calcining the ore in open piles the chert would pop out, and leave the ore pure. Brown ore, he went on to say, was most peculiar in that respect. It might contain a good deal of chert, but when it was heated the chert would spring away from the ore, and it was dangerous to stand near the pile. They all moved back, and the orator proceeded! The method of improving cherty brown ore by popping the chert out may be patentable, but is not in use in this State, or elsewhere. Attention is being drawn more and more to calcining in gas-fired kilns, and of the various kinds the Davis-Colby is preferred. In this kiln the current of heated gas and flame is drawn across the ore as it descends between the outer walls of the combustion chamber and a central space connected with the stack. The kiln is built of any convenient size, from 100 to 150 tons capacity, and is fired with producer gas.

Allowing 7 per cent. of hygroscopic water, removable at 212 deg. F., and 7 per cent of combined water, removable only at red heat, a kiln holding 125-140 tons of raw

THE LIMONITE, OR SO-CALLED BROWN ORES.   53

ore will deliver from 107 to 120 tons of thoroughly and uniformly calcined ore per 24 hours, with a consumption of 2¼ to 3 tons of coal. To calcine one ton of raw ore (2240 lbs.) requires about 52 lbs. of coal

The advantages of the gas-fired kiln are economy of labor, and uniformity of product. These advantages maintain under all conditions, except where the price of coal is prohibitory, and even there the wood-fired or charcoal-fired producer may be used.

The use of all brown ore in coke furnaces may be rendered necessary by contracts specifying that the iron shall be made from brown ore, or by proximity to deposits known to be very considerable. A determination on the part of furnace owners to make a special high grade charcoal iron would also entail the exclusive use of brown ore.

A kiln to treat 140 tons of raw ore per day, with producer and all necessary fittings, will cost about $7,000, and will yield ordinarily about 120 tons of calcined ore. This amount would contain from 60 to 65 tons of iron, and would be equivalent to 20 per cent. of the ore burden for 2 150 ton furnaces.

The freight on a ton of raw ore from the washer to the furnace may be taken at 25 cts. in the Birmingham district, and if the ore averages 47 per cent. of iron we would have 1952.8 lbs. of iron costing for freight 25 cts.

The freight on a ton of calcined ore would also be 25 cents, but it would contain 54 per cent. of iron, or in the ton 1209.6 lbs. of iron. So far, therefore, as concerns the transportation charges we would get 1209.6 lbs. of iron in the calcined ore at the same price paid for 1052.8 lbs. in the raw ore. Each ton of calcined ore delivered at the furnace would contain 156.8 lbs. of iron more than a ton of raw ore. If it requires 4 men in the stockhouse, as bottom-fillers, to handle 140 tons of raw ore per day, containing 65.8 tons of iron, 3 men could handle the

121.7 tons of calcined ore required for the same amount of metal. So far as concerns the handling of the ore in the stockhouse there would be a saving of one man at each furnace by substituting calcined ore for raw ore.

The economy becomes even more striking if we consider the kiln as situated at the furnace, so that the bottom-fillers could draw the ore from the shutes. At one well managed plant this has been the practice for several years. The trams come in from the washer and discharge into the kiln. The bottom-fillers draw from the shutes into the buggies, and the hot one goes at once to the furnace. At this establishment it has been shown that there is great advantage in the use of calcined ore, irrespective of the easy way of handling it in use, and it fortunately happens that it is able to compare, for a term of years, the practice on raw ore, pile-calcined, and kiln-calcined ore.

It is not going too far to say that it would be profitable to erect kilns at the furnaces, even when the ore has to be hauled at a freight cost of 25 cts. per ton, or even more.

Excessive freight charges on ore would, of course, militate against this proposition, but until they rise beyond 40 cts. per ton calcining would be advantageous.

The erection of kilns at the mines, except under unusual conditions, can not be recommended, for the reason that the life of a brown ore deposit is uncertain.

But at the furnace, and especially where coke is made on the spot and it is possible to calcine with waste gases from the ovens, this objection is removed. The furnace operator would be able to buy ore from the smaller mines which can not incur the expense of building kilns, the entire process would be under one management, and the utilization of gases now going to waste would, of itself, show a profit.

It is a truth of general application that it pays to calcine brown ore, for it has been shown to be beneficial wherever it has been carefully and faithfully carried out.

MILL CINDER.

Another material used in the Birmingham district, as a source of iron, is mill cinder.

It is a product from puddling furnaces, and is worth from 90 cents to $1.00 a ton, delivered.

The composition varies somewhat, as the following analyses show:

Equal parts, by weight, of heating furnace and puddle cinder; metallic iron, 56.59 per cent.

Equal parts, by weight, of cinder made with coal, cinder made with gas, and puddle cinder; metallic iron 51.33 per cent.

Equal parts, by weight, of flue and tap cinder; metallic iron, 50.08 per cent.

The average composition of ordinary mill cinder is about as follows:

	Per cent.
Metallic iron	50.00
Silica	20.00
Alumina	1.50
Lime	0.50
Sulphur	1.50
Phosphorus	0.60

It is not used regularly, but in broken doses, as a "scouring material."

BLUE BILLY, PURPLE ORE.

Residue from pyrite burners in sulphuric acid works.

This material is occasionally used, being purchased from the sulphuric acid factories in Atlanta, Pensacola, &c. It carries generally more than 60 per cent. of iron, but the content of sulphur is quite variable and may be as much as 2.50 per cent. Properly roasted, i. e., with sulphur below 0.50 per cent., it would commend itself as a source of iron.

CHAPTER III.

THE FLUXES.

The material used for flux in the state is either limestone, dolomite, or a mixture of the two in varying proportions. It is now very largely sold on analysis, samples being drawn from each car received. The basis of sale is the percentage of silica, some of the contracts starting at 2.50 per cent. and others at 3.50 per cent. When the stone is sold on analysis it is customary to employ a sliding scale, as has already been explained under the brown ore. Suppose the base is 3.50 per cent. of silica. The scale is so arranged that for each quarter of one per cent. above 3.50 per cent., two-tenths of a cent per ton is taken off, and for quarter of one per cent. below 3.50 per cent. of silica two-tenths of a cent is added. Thus if the delivery price is 60 cents per ton for a 3.50 per cent. stone, and the silica should run to 3.75 per cent., the price would be 59.8 cents per ton, and if the silica should fall to 3.25 per cent., the price would be 60.2 cents per ton. If the silica should rise to 5 per cent. the price per ton would be 68.8 cents, and if it should fall to 2.00 per cent. the price would be 61 cents.

The average analysis of the limestone used in the state may be stated as follows:

Silica 4.00%
Oxide of iron and alumina. 1.00
Carbonate of lime...... 94.60   Lime 53.00%

It not infrequently happens that the stone is much higher in silica than this average. Instances are on record in which the silica was 8.00 per cent. In such cases the production of iron is attended with considerably higher cost than when the better stone is used.

Limestone was the only flux used up to within the last few years. Since that time the use of dolomite has largely increased, the great advance being within the last year. In the manufacture of basic iron 'intended' for the open hearth steel furnace it was soon found that the use of dolomite was a decided advantage, especially in the elimination of sulphur. Whether this result was due to the fact that the dolomite carried only 1.25–1.50 per cent. of silica as against 4.00 for the limestone, or whether the presence of magnesia was of real benefit, so far as concerns the elimination of the sulphur, is still in dispute. The fact, however, remains that in the production of basic iron, sold on analysis under severe restrictions as to quality, only dolomite is used. Aside from its low silica content, the dolomite possesses the further advantage of great uniformity of composition. This is a point very much in its favor. My own experince with limestone in this state covers something like 22,000 cars, and with dolomite about 2,500 cars. The former is subject to considerable variation in respect to silica, while the latter, in so far at least as concerns the lump stone, is of remarkable uniformity. The highest amount of silica observed in the lump dolomite is a trifle over 1.50 per cent., the ordinary range being from 0.75 to 1.25 per cent.

Extensive deposits of both limestone and dolomite exist within eight miles of Birmingham. The haul for limestone is, however, about thirty miles, only the dolomite being worked within the immediate vicinity. So far as my observation goes, the average composition of the dolomite used may be taken as follows:

Silica .... 1.50 %
Oxide of iron and alumina 1.00 %

THE FLUXES.

Carbonate of lime......54.00% Lime 30.31%
Carbonate of magnesia..43.00% Magnesia 20.71%

The proportion between the magnesia and the lime does not vary much from 1:1.50.

Both the limestone and the dolomite carry small amounts of sulphur, the maximum so far observed being 0.11 per cent.

As in the limestone quarries there are layers of siliceous material interfering with the quality of the material, so in the dolomite quarries there are ledges of almost pure silica, white as porcelain. They seem to be flinty concretions occurring in more or less regular bands, from one half an inch to three inches in thickness. It is customary to separate these flinty nodules from the stone by hand before it is shipped. They do not seriously interfere with the quality of the dolomite if care is used in the separation. Otherwise they are extremely objectionable.

The impure limestone is of a much darker color than the good stone, but the impure dolomite is generally much lighter in color than the remaining portion. There is a kind of dolomite that occurs in some of the quarries that is very deceptive to the eye. It looks not unlike coarse brown sugar, has the same damp appearance and glistens in the sunlight. To the hand it feels sandy, but on analysis it is found generally to be the best stone in the quarry. Some samples have given only 0.25 per cent. of silica. Not all of this loose, sandy looking dolomite is good, however, for it sometimes happens that it carries more than 3.00 per cent. of silica, and one sample was found to contain nearly 4.00 per cent. It does not form a large proportion of the material in the quarry, and is mined and shipped with the other stone.

Both the limestone and the dolomite are quarried on the face, no underground work being required. Crushed stone or lump is shipped as occasion may demand.

The amount of stone used per ton of iron varies, of course with the quality of the stone, with the nature of the ore and fuel, and, to some extent, with the grade of the iron required. The range is from 0.30 ton to 0.80. This subject will be discussed in the chapter on Furnace Burdens, which will be devoted to the general practice throughout the state, different types of burdens being selected with reference to the consumption of raw materials per ton of iron and the cost of the same.

No attempt has been made on any considerable scale to use calcined stone, whether limestone or dolomite, except in so far as the calcination of hard ore may be considered as an attempt to calcine the carbonate of lime contained in it.

It is necessary here merely to state the question in general terms. As has been already remarked, in the discussion of the hard ore, we have in this State an intimate mixture of oxide of iron, silica and carbonate of lime. The best of it contains on the average 37 per cent. of iron, 13.44 per cent. of silica, and 15.45 per cent. of lime as carbonate. The admixture of these materials is far more perfect than could be attained by any practical mechanical means, although some of the ore is not self fluxing. This being the case we can ask ourselves if it is more economical to employ this ore, in which the flux is already so well mixed with the silica, than to use an ore of far less content of lime and therefore requiring the addition of flux. At the first glance it would appear that it is better to avail ones self of whatever advantages nature herself has conferred upon us in the way of an ore carrying its own lime. But the matter can not be settled out of hand and without careful investigation of all the data bearing upon it. From the standpoint of the furnace man, if he could depend on securing self fluxing ore regularly, the matter resolves itself into the simple consideration as to whether he can make as much iron and as cheap iron in the one way as in the other. He may, indeed, go a step farther and ask if he makes iron more cheaply in the one way than in the other. Having settled this, he has no further concern with the matter. If he can make iron more cheaply by using a greater and greater proportion of hard ore than by using an ore which requires the addition of extraneous flux, it is his duty to do it. This, however, is a one-sided view. There are other investments in the State that must be regarded as well as investments in furnaces. How is it with the contractor for ore and flux? Would his business be hindered by the substitution of hard ore for stone? If his profit on the ore was the same as his profit on the stone, no great hardship would follow the increase in the use of the one and the decrease in the use of the other. But if it should happen that his profit in mining stone was greater than his profit in mining hard ore, and there should be such an increase in the consumption of hard ore as to destroy the value of his stone quarry, he would not be apt to appreciate the advantages of the change. In this respect this iron district differs from any other in the country, and the relations of stone to ore burden vary perhaps more widely than elsewhere. The ability of the furnaces to diminish at will the consumption of limestone, places them in a very independent position. If the price of stone is too high, they can run on increased proportions of hard ore. If they succeed in obtaining the stone at reasonable cost, they take off hard ore and put on soft or brown. For instance, a certain coke furnace during a certain month last year made about 5,000 tons of iron with an ore burden composed of 50.9 per cent. hard, and 49.1 per cent. soft ore. The total burden was as follows:

Hard ore................ 27.7 per cent.
Soft ore................ 26.7 "
Limestone............... 15.5 "
Coke .... 30.1 "

100.00 "

The consumption per ton of iron was:
Ore ....2.36 tons (2240 lbs.)
Stone ....0.67 "
Coke ....1.32 "

4.54

And the cost per ton of iron was:
Ore............................$1.32
Stone.......................... 0.34
Coke 1.83

$3.49

The consumption of coke per pound of iron made was 1.32 lbs., and practically all of the iron was of foundry grades.

Shortly before, the same furnace was running on 33.4 per cent. hard, 65.3 per cent. soft, and 1.3 per cent. brown ore. The total burden was:

Hard................... 17.0 per cent.
Soft................... 33.1 "
Brown.................. 0.6 "
Limestone.............. 16.9 "
Coke .... 32.4 "

100 "

The consumption per ton of iron, of which something over 4,600 tons were made, was, in tons of 2,240 lbs.:
Ore............................ 2.20
Limestone...................... 0.73
Coke 1.41

4.34

The cost per ton of iron was:

Ore	$1.26
Stone	0.43
Coke	1.83
	$3.52

The consumption of coke per pound of iron was 1.41 lbs., and in this case also practically all of the iron made was of foundry grades. In these two cases there was a saving of nine cents per ton of iron by increasing the proportion of hard ore and lessening the amount of limestone added. The ore cost six cents a ton of iron more than when the larger proportion of soft ore was used, so that the net gain was three cents per ton of iron, $3.49 for the hard ore burden, and $3.52 for the other.

But with the lesser amount of hard ore the furnace made 358 tons of iron more than with the greater amount. This has to be set to the credit of the soft ore burden.

Perhaps no positive conclusions can be drawn from one or two instances, and as the whole matter will be fully discussed under Furnace Burdens, it may be best to defer any further remarks.

Enough, however, has been said in this chapter on the fluxes to direct attention to the importance of the considerations advanced. The future of the iron industry in the State depends not on any one circumstance or condition, howsoever vital it may seem, but upon the resultant of a number of forces, some of whose effects may be at the present but dimly foreseen. It is possible that the relation between hard ore and limestone, or dolomite, is one of these.

Dolomite as a Flux for Blast Furnace Use,

BY

Ed. A. Uchling,

(*Proc. Ala. Indus. & Sci. Soc.*, Vol. IV, 1894, p. 24.)

Dolomite is the name given in honor of the French geologist, Deodat-Guy-Silvain-Tancrede Gratet de Dolomieu, to a carbonate of lime and magnesia in which these two constituents occur in equal or nearly equal equivalents.

The atomic weight of magnesium is 24, while that of calcium is 40; but as each of these atoms is combined, respectively, with an atom of oxygen to a molecule of the oxide, and each respective molecule of the oxide is combined with a molecule of carbonic acid to form the carbonate, and as the molecular weight of the carbonic acid is 44 in each case, it follows that an equivalent of carbonate of magnesia will weigh 84, while one of carbonate of lime will weigh 100. In fluxing power, i. e., in the power to combine with silica and form a fusible slag, these equivalents are equal, because the power of a base to combine with an acid does not depend upon its atomic weight, but upon its chemical affinity, from which it further follows that 84 parts, by weight, of magnesia have the same value as a flux as 160 parts of lime.

Pure dolomite is, in round numbers, composed of 46 per cent. of carbonate of magnesia and 54 per cent. of carbonate of lime. Now, because the fluxing power, as shown above, is equal, equivalent for equivalent, and because there are as many equivalents of magnesia in the 46 per cent. as there are equivalents of carbonate of lime in the 54 per cent., it follows that 100 pounds of pure dolomite are equal to 108 pounds of pure limestone in fluxing power.

The dolomite which is available in the Birmingham

THE FLUXES. 65

district is of exceptional purity, both as to the foreign matter it contains and as to the proportion of lime and magnesia carbonate of which it is composed, viz : 55 per cent. of the former and 43 per cent. of the latter, with only 2 per cent. of foreign matter. The theoretically pure dolomite should be composed of 45.65 per cent. of carbonate of magnesia and 54.35 per cent. of carbonate of lime.

The limestone of the district is vastly more irregular. While there are some ledges of exceptional purity, there are others that are entirely worthless for fluxing purposes. The worst feature of these irregularities is that the impure ledges make their appearance in all the quarries thus far opened. For this reason it has not been possible to get limestone that will average above 96 per cent. of carbonate of lime, and 94 to 92 and even down to 90 per cent. is not infrequently the average of whole shipments.

We will take for granted that with the exercise of sufficient care in the quarries, a limestone of an average of not to exceed 4 per cent. of impurities can be furnished.

In determing the value of a stone as a flux, it is not only necessary to deduct the impurities it contains, but in addition to that, as much of the base as is necessary to flux these impurities. What remains only can be considered as available flux, and has value in the blast furnace. To get at the available flux, we must deduct 2 per cent. from the carbonate of lime for each unit per cent. impurity in the stone. Taking the limestone at 96 per cent. of carbonate of lime and deducting from this 8 per cent. to take care of its own impurities, we have left for available flux 88 per cent. of carbonate of lime.

As the average dolomite contains only 2 per cent. of impurities and 43 per cent. of carbonate of magnesia with 55 per cent. of carbonate of lime, we will have, after deducting 4 per cent. from the carbonate of lime, 51 per cent. of this material, and 43 per cent. of carbonate of magnesia. Reducing the carbonate of magnesia to its equivalent in fluxing power of carbonate of lime, we have, because the fluxing powers of the two carbonates are to each other as 84 to 100,

$$\frac{43 \times 100}{84} \times 51 = 102.19.$$

The relative values of the two available fluxing materials of the district are, therefore, to each other as 88 is to 102.19. That means that 88 tons of dolomite will do as much work in the blast furnace as 102.19 tons of limestone. Put into dollars and cents, this means that if dolomite can be bought for 60 cents a ton, limestone is worth only 52 cents a ton; or if limestone costs 60 cents, dolomite is worth 69.5 cents a ton.

There is only one valid objection that can be brought up against the use of dolomite as a flux in the blast furnaces, and that is that magnesium has less affinity for sulphur than calcium has, and dolomite is therefore less efficient as a desulphurizer than limestone, to the extent that caustic lime is displaced by magnesia.

This objection, however, becomes quite insignificant where the ores are free from sulphur, as is the case in the Birmingham district. When a considerable proportion of hard ore is used in the mixture, its lime, in connection with what is contained in the dolomite itself, is ample to take care of the sulphur contained in the coke.

One-quarter to one-half dolomite has been regularly used in the Sloss furnaces for nearly two years, and, at intervals, as high as three-fourths have been put on with the best results. The ore mixture being half hard and half Irondale (soft) at the city furnaces, and from one-fourth to one-third brown with generally equal proportions of Irondale (soft), and hard at the North Birmingham furnaces.

The coke used contained considerably above the average amount of sulphur found in the average coke of the district.

The iron was of as good quality as could have been produced with all limestone as a flux, and the furnaces have worked more regularly than they did prior to the use of dolomite. The assertion that the use of dolomite has a tendency to make light colored iron is not sustained by fact. Some of the most celebrated foundry irons are made with all dolomite as a flux. The writer had used it for years, while in charge of the blast furnaces of the Bethlehem Iron Company, prior to coming down here, and experienced no difficulty in keeping the sulphur within the required limits, even with ores containing as high as 1.5 per cent. of that element.

The Illinois Steel Co. are also using dolomite exclusively in their Joliet Works. They are doing very good work, and have no trouble with the sulphur whatever.

The deficiency of dolomite to carry off sulphur is probably very much exagerated. There are impure dolomites as well as impure limestones; but when of good quality and used intelligently and without prejudice, it always gives good satisfaction. In addition to its superior fluxing power there is decidedly less tendency to 'hanging' with dolomite than with carbonate of lime.

To Mr. C. A. Meissner belongs the credit of having first systematically tried dolomite with the Birmingham ores.''

THE FUELS.

The fuel used in the blast furnaces of the state is coke and charcoal, no coal being used. There are no known seams of coal that could be used without coking, as is done in Ohio in this country, and in Scotland, particularly, abroad.

Coke.

There is, perhaps, no subject connected with the iron business that gives rise to more discussion than that of coke. There are so many different kinds made, and so great diversity among them in respect of chemical and physical properties, that it is almost a hopeless task to attempt to set the matter forward in a manner satisfactory to all concerned. Even in this State, which produces about 10 per cent. of the coke made in the United States, there is a very considerable difference in quality between the various grades of this fuel.

This chapter is not a treatise on coke, nor is it necessary to enter upon the subject beyond what is required to explain the situation in the State.

Three kinds of coke are made here, from lump coal, run of mines, and washed slack, and each of these three may be 48 hr. or 72 hr. coke. Regarded in this way, and excluding mixtures, of which there may be endless variety, we have six different kinds to-wit:

48 hour—	72 hour—
Lump,	Lump,
Run of mines,	Run of mines,
Washed slack,	Washed slack.

The ordinary practice is to use 48 hr. coke, and perhaps 90 per cent. of the coke is of this kind. The chief difference between the 48 hr. coke and the 72 hr. coke is in the strength, or the ability to resist abrasion and crushing, the latter having somewhat the advantage in this respect.

The following table gives the results of some experiments undertaken to establish the crushing strain of a number of different cokes made in Alabama, together with the analysis of the samples.

It will be seen that the 72 hr. is a good deal stronger than the 48 hr. coke made from the same coal. The table is taken from the writer's article in the Proc. Ala. Industrial and Scientific Society, 1892, Vol. I, p. 17:

TABLE III.—ALABAMA COKES.

	Proximate Analysis.			Ultimate Anal.				Color of Ash.	Crushing Strain Pounds Per Square Inch.
	Moisture.	Vol. and Combust. Matter.	Fixed Carbon.	Ash.	Sulphur.	Carbon.	Hydrogen.	Oxygen.		
Black Creek	0.50	0.90	94.90	3.90	0.79	84.87	5.52	4.62	Reddish brown	Crack'd at 400, broke at 1000
Blocton	0.10	0.60	93.20	6.10	1.05	83.18	6.43	3.14	"	Broke suddenly at 445.
Blue Creek, 48 hr	0.40	0.70	86.00	12.90	1.23	76.23	6.18	3.06	"	Cracked at 486, broke at 750
Blue Creek 72 hr	0.05	0.90	87.25	11.80	0.99	78.27	7.38	1.51	"	Crack'd at 720, broke at 1100
Coalburg	0.70	2.55	81.15	15.60	2.21	75.25	4.25	1.99		Broke suddenly at 550.
Mary Lee	0.90	3.10	86.45	9.55	0.93	80.32	5.65	2.65	Buff	Broke at 488.
Pratt	0.90	1.40	89.80	7.90	0.82	77.86	7.10	5.42	Reddish brown	Cracked at 350, broke at 540
Pratt, black ends	0.25	0.85	85.45	13.40	1.65	74.61	6.07	3.97	"	Broke at 300.
Standard (Brookwood)	0.20	1.10	82.15	16.55	1.43	69.18	5.57	7.07	Buff	Broke suddenly at 320.
Standard (Milldale)	3.05	2.50	86.80	7.65	1.47	76.63	3.56	7.64	"	Broke gradually at 545.
Standard (Mixture)	0.30	1.00	82.10	16.60	1.49	71.25	4.70	5.66		Broke gradually at 800.
Gas Carbon (Birmingham)	0.20	0.70	93.20	5.90	1.23	84.27	4.61	3.79	Reddish brown	Broke at 600.
Gas Carbon (Mobile)	2.90	3.25	83.05	10.80	1.45	71.74	7.40	5.71	Dark gray	Broke at 700.
" Thomas Ovens.										

THE FUELS. 71

Some work has been done in the direction of determining the apparent and the true specific gravity, the percentage of cells by volume, and the volume of cells in 100 parts, by weight, of the coke. The investigations are still in progress, but it may be as well to give here the results already reached. They are averages of a considerable number of determinations, and the coke was in every case 48 hr. bee-hive; coal not crushed:

Coke—	App. sp. gr.	True sp. gr.	% of cells by Vol.	Vol. of cells in 100 pts.	Ash.
Washed slack,	0.861	1.784	51.69	60.24	10.82
Run of mines, same coal,	0.860	1.829	52.88	61.35	15.00

There does not appear to be any regularity in the relation between the ash and the percentage of cells by volume, or the volume of cells in 100 parts, by weight, of the coke, basing this opinion on some 30 determinations. A larger number may modify this conclusion.

In regard to another important quality of coke, viz., its ability to resist, at red heat, the dissolving action of carbonic acid, very little information has been gathered. It must be remembered that the scientific study of the materials used for iron making in this state is still in its infancy. The routine furnace work precludes, to a great extent, any excursions into fields which, although attractive, do not seem to those who pay the bills sufficiently promising for immediate cultivation. What has been done in the last few years is, however, very encouraging, and we are constrained to hope that the next few years will witness the extension of scientific investigations in many directions at present closed.

The average composition of the coke used in the blast furnaces may be stated as follows:

Coke from Run of Mines Coal.

	PER CENT.
Moisture	0.75
Volatite and combustible matter	0.75
Fired carbon	84.50
Ash	14.00
	100.00

Sulphur.................0.90—1.60 per cent.

Coke from Washed Slack.

Moisture	0.75
Volatile and combustible matter	0.75
Fixed carbon	88.50
Ash	10.00
	100.00

Sulphur.................0.80—1.10 per cent.

Coke from Lump Coal.

Moisture	0.75
Volatile and combustible matter	0.75
Fixed carbon	87.00
Ash	11.50
	100.00

Sulphur.................1.00—1.30 per cent.

In chemical composition there does not seem to be any material difference between the 48 hr. and the 72 hr. coke.

The composition of the ash of the various cokes in use may be given as follows:

Run of Mines.

	PER CENT.
Silica	47.03

Ferric oxide..................................12.46
Alumina......................................33.62
Lime... 1.53
Magnesia..................................... 1.69
Sulphur...................................... 0.75

Washed Slack.

Silica.......................................45.10
Ferric oxide.................................12.32
Alumina......................................31.60
Lime... 1.50
Magnesia.....................................Trace.
Sulphur...................................... 0.50

Lump.

Silica.......................................46.00
Ferric oxide.................................12.00
Alumina......................................32.00
Lime... 1.00
Magnesia..................................... 0.50
Sulphur...................................... 0.60

It would be interesting to know if the amount of ash and its composition influenced the strength of the coke, or whether the treatment of the coal, prior to charging the ovens, and the duration and temperature of the process should alone be looked to in explanation of this point.

It does not seem probable that the amount of ash or its composition, *per se*, would influence the strength of the coke as much as the distribution of the ash constituents in the coal.

That is, if the coal was finely pulverized before charging there would be a more equable distribution of the ash-constituents with consequent uniformity of composition in the coke. But uniformity of composition, however desirable, does not necessarily imply increase in strength. Granting that there would be increase in strength is this effect beneficial when the coke is already strong enough? If the coke made from any coal, without pulverizing, were already strong enough, the only advantage in pulverizing would be in the greater uniformity of composition. But some coals do not yield strong coke unless they are pulverized. Whether this is due to the irregularity of the distribution of the ash, or the bituminous matter, or the relation between the coking and the non-coking constituents of the coal, is not known. When, however, such coals are pulverized they often make excellent coke.

The composition of the ash of coke, by affecting its fusibility, may affect also its strength, the size and shape of the cells and the thickness of the cell walls. But of such matters very little is known.

It requires a great deal of time to make such investigations, as well as skill and perseverance.

The composition of the ash of coal, whatever effect it may have on the quality of the coke made from it, cartainly has an important bearing on furnace practice. It must influence the fusibility of the burden, and to a greater or lesser degree affect the consumption of limestone, whether this be the carbonate of lime in the hard ore, or extra stone. The more acid the ash the more base is required for fluxing.

The amount of coke used per ton of iron varies, of course, with the nature of the coke, and of the other constituents of the burden; with the kind of iron made, the shape and size of the furnace, the rate of driving, and other circumstances grouped generally under the term "furnace practice". The range is from 1.16 to 1.72 tons of 2240 pounds. From an examination of 150,000 tons of iron made from 1890 to 1895 under varying conditions the lowest consumption for a period of one month was 1.16 tons per ton of iron. In this particular case the furnace was working on all brown ore, the burden being composed of brown ore 52.9, limestone 20.4, and coke 26.7. The tons of iron made per charge was 1.53 tons, number of charges 1802, total iron made 2766 tons, of which 99.1 per cent. was of foundry grades. The consumption of materials per ton of iron made was ore 2.31 tons, stone 0.89 ton, and coke 1.16. For further information regarding this case reference may be made to No. 1 table VII page 94.

The particular case in which 1.72 tons of coke were used per ton of iron made was when a furnace was running on the following mixture, stated as per centages, hard ore 53.7, soft ore 34.2, brown ore 12.1. The entire burden was composed as follows in percentages, hard ore 28.5; soft ore 18.2, brown ore 6.3, limestone 10.6, coke 36.4. The iron made per charge was 1.88 tons, number of charges 1819, total iron made 3418 tons, of which 92 per cent. was of foundry grades. The consumption of material in tons per ton of iron was as follows:

Ore..2.51
Stone..0.49
Coke..1.71

For further information see No. 22, table VII, page 94.

The average consumption of coke per ton of iron may be taken at 1.41 tons of 2240 pounds. This would mean that for producing the 835,851 tons of coke iron in 1895 there were used 1,179,375 tons of coke and that 250,000 tons of coke made in the State during that year were diverted to some other purpose.

The average for the best coke made in the State may be taken at 1.30 tons of 2240 pounds for a ton of iron of 2240 pounds. A pound of iron has been made in the State with less than a pound of coke, but for a very limited period.

This matter will be taken up more fully in the chapter on Furnace Burdens, as tables have been prepared based on more than 83,000 charges and an iron production of nearly 150,000 tons over a period of several years.

There has been a notable decrease in the consumption of coke per ton of iron since the introduction of coke made from washed slack coal. It is much superior to ordinary coke both in structure and composition, and might be still further improved by pulverizing the coal before charging the oven, as in this way a better distribution of the ash is rendered possible as well as a stronger coke.

No constituent of the burden responds as readily to variations in furnace practice as coke. It forms generally more than a third of the burden, and always more than half of the total cost of the materials entering into a ton of iron is chargeable to coke. It is not only the most costly single ingredient it is more costly than the ore and the stone taken together.

Economy in the use of coke is, therefore, the most important economy that can be set on foot and carried out in connection with the manufacture of pig iron in this state. Better ore and better stone are needed if there is to be no better coke. To improve the ore and the stone is to increase the yield of iron per charge, and to decrease the consumption of the most costly material entering the furnace, i. e. coke.

The following table gives a bird's eye view of the coke industry in Alabama from 1880 to the close of 1895, and is compiled from the reports of Joseph D. Weeks to the United States Geological Survey, Div. Mineral Resources,

TABLE IV.

Year.	Establishments.	Ovens. Built.	Ovens. Building.	Coal Used, Tons.	Coke Produced, Tons.	Yield of Coal in Coke per ct.	Value of Coke. Total.	Value of Coke. Per ton.
1880	4	316	100	106,283	60,781	57	$ 183,063	3.01
1881	4	416	120	184,881	109,033	59	326,819	3.00
1882	5	536	.....	261,839	152,940	58	425,940	2.79
1883	6	767	122	359,699	217,531	60	598,473	2.75
1884	8	976	242	413,184	244,009	60	609,185	2.50
1885	11	1,075	16	507,934	301,180	59	755,645	2.50
1886	14	1,301	1,012	635,120	375,054	59	993,302	2.65
1887	15	1,555	1,362	550,047	325,020	59	775,090	2.39
1888	18	2,475	406	848,608	508,511	60	1,189,579	2.34
1889	19	3,944	427	1,746,277	1,030,510	59	2,372,417	2.30
1890	20	4,805	371	1,809,964	1,072,942	59	2,589,447	2.41
1891	21	5,068	50	2,144,277	1,282,496	60	2,986,242	2.33
1892	20	5,320	90	2,585,966	1,501,571	58	3,464,623	2.31
1893	23	5,548	60	2,015,398	1,168,085	58	2,648,632	2.27
1894	22	5,551	50	1,574,245	923,817	58.7	1,871,348	2.25
1895	..	5,658	.....	2,459,465	1,444,339	58.7	3,033,521	2.10

The average value of the coal used in making coke in 1895 was 87½ cents per ton.

A few years ago it was customary to use run of mines coal for coking, but since 1892 the tendency has been towards slack, both unwashed and washed.

To show the changes that have come about during the last few years the following table, taken from the same authority, is given here:

TABLE V.

Character of Coal used in making Coke in Alabama.

Year	Run of Mine — Unwashed Tons	Per cent	Run of Mine — Washed Tons	Per cent	Slack — Unwashed Tons	Per cent	Slack — Washed Tons	Per cent	Total
1890	1,480,669	81.8			206,106	11.3	123,189	6.9	1,809,964
1891	1,943,469	90.6			192,238	9.0	8,570	0.4	2,144,277
1892	2,463,366	95.3			11,100	0.4	111,500	4.3	2,585,966
1893	1,246,307	61.8	51,163	2.5	292,198	14.6	425,730	21.1	2,015,398
1894	411,097	26.1	7,429	0.5	477,820	30.3	677,899	43.1	1,574,245
1895	1,208,020	49.1			32,068	1.3	1,219,377	49.6	2,459,465

The substitution of washed slack for unwashed run of mines coal in coke making is very evident. Even so late as 1892 more than 95 per cent. of the coal sent to the ovens was unwashed run of mines, a little over 4 per cent., on the other hand, being washed slack. In 1894, the percentages were 26.1 and 43.1, and in 1895 49.1 per cent. and 49.6 per cent.

The use of washed slack enables the mine owners to avail themselves of what would otherwise be of little value, and to make a better coke of this material than is made of run of mines coal.

FURNACE BURDENS.

Few considerations affecting the production of pig iron are of more importance than the proper admixture of the materials from which the iron is made. Pig iron is made from iron ore, coke, charcoal or other fuel, and limestone, or dolomite for a flux. The ore contains the iron mixed with various substances from which by process of reduction, the iron is freed. Iron does not exist in ore as such, but is combined, generally, with oxygen, and mixed with siliceous matter. To remove the oxygen some form of carbon is used, such as coke, charcoal, anthracite coal, or a kind of bituminous coal known as splint coal. To remove the siliceous (sandy) matter carbonite of lime (limestone) is used, or a mixture of carbonate of lime and carbonate of magnesia (dolomite). These materials, the ore, the fuel and the stone, are melted in the blast furnace, and there are obtained from them pig iron and slag, or cinder.

Coke Furnaces.

The largest furnaces in Alabama are 80 feet high, and 19 feet 6 inches wide in the bosh, or widest part. The greatest amount of pig iron ever made in a furnace in one day in this State was 265 tons, and for its production there were required 588 tons of ore, 62 tons of limestone and 265 tons of coke, all of 2,240 lbs.

It is by no means unusual for a furnace to make 200 tons of iron a day, and for this there would be required 480 tons of ore, 280 tons of coke, and 25 tons of stone, if the proper amount of hard ore were used. The average number of tons of material handled per ton of iron made is about 4.44 in coke furnaces, so that for the 835,851 tons of coke pig iron made in 1895 there were handled 3,711,178 tons of material, of which 2,089,627 tons were ore, 442,176 tons were stone (limestone and dolomite), and 1,179,375 tons were coke. These are approximate figures. The amount of ore required to make a ton of iron varies from 2.10 tons to 2.87 tons, the average being close to 2.50. The average amount of coke used per ton of iron made is 1.41 tons of 2240 lbs., the range being from 1.16 to 1.60.

The average amount of stone used per ton of iron made is about 0.53 ton, the range being from 0.10 to 0.88.

The amount of each material entering the furnace per

day is not a matter of guess, or of indifference, but is carefully determined from the chemical analysis. It is customary to fill the furnace and keep filling it by "charges," each "charge" being composed for the most part of ore, coke and stone. Thus, for instance, a "charge" may be composed of 5,600 lbs. of coke, 10,080 lbs. of hard ore, 2,740 lbs. of soft ore, and 620 lbs. of limestone, and the furnace will take from 80 to 90 charges per day, and should yield 200 tons of iron. The proportion between the various elements of the charge, as well as the total weight of the charge, and the number of charges per day, are all subject to change, but unless there is urgent necessity the daily alterations should be very slight. Having once established the proper burden, it is not advisable to change it, nor is it necessary to do so if the materials can be provided in sufficient quantity and with sufficient regularity, and uniformity of composition. But changes of burden are very frequently made, so frequently in fact that the necessity for them constitutes the greatest obstacle in the path of successful furnace management in this state. It is the lion in the way, unchained at that. In comparing furnace practice in Alabama with furnace practice in Pennsylvania, for instance, one is impressed at the outset with the frequent and in many cases violent changes in the burden in the first place, and in the second with the large tonnage handled per ton of iron. This tonnage is referrable to the raw materials going into the furnace, and to the cinder which, of course, has to be removed. This condition of affairs will remain as it is now until better ore can be obtained, as the ore comprises about 56 per cent. by weight of the burden, being more than the stone and the fuel together, and is subject to wider variations in physical and chemical composition than either the stone or the fuel.

In discussing furnace burdens, therefore, it must be

understood that we do so with some reservations. To present the matter briefly and in a general way, as becomes the character of this publication, and yet truthfully as far as we shall go, is difficult. Generalizations can be accepted only with the grain of salt, and should be based on a certain set of conditions. Given these we may derive valuable information, but to utilize them to the best advantage one must know more than appears on the surface.

It may be advisable to take up the subject first from the standpoint of the coke furnace, and then discuss, briefly, the charcoal practice.

We will divide the coke practice into two main heads:

1st. Burdens composed, so far as concerns the ore, of hard ore and soft ore, the proportion of the hard ore rising from 48.2 per cent. to 100 per cent.

2d. Burdens composed, so far as concerns the ore, of hard ore, soft ore, and brown ore, the proportion of brown ore rising from 1.30 to 100 per cent.

1st. Burdens composed, so far as concerns the ore, of hard ore and soft ore, the proportion of hard ore rising from 48.2 per cent. to 100 per cent.

In order that the same basis of comparison may be used, we have taken the delivery prices of the raw materials as follows:

    Per ton of 2,240 lbs.
    Hard ore...............67.5 cts. per ton.
    Soft ore................55.4 " "
    Limestone..............63.4 " "
    Coke..................$1.75 "

These prices are very close to the averages for shipments during 1895.

The table that has been prepared is based on actual furnace records, and comprises results obtained from the examination of 32,917 charges, the amount of pig iron represented being 50,360 tons. The years selected were 1889, 1890, 1893, 1894 and 1895. The tons referred to are of 2,240 lbs. The table includes the year, the private number, the number of monthly charges, the percentage composition of the ore burden and of the total burden; the iron made, per charge, and for each month, and the percentage of foundry grades (including F. F. or 4 F., but excluding Gray Forge, mottled and white); the consumption of ore, stone and coke in tons per ton of iron made; the cost of the ore, the stone and the coke per ton of iron; the percentage distribution of this cost; and the pounds of coke required to make a pound of iron. The calculations have been somewhat laborious but the results are extremely interesting and important. They do not cover as much ground as could be wished, but the pressure of other matters compelled an abridgement of the original plan.

We will give a table of results from the same furnaces, consecutive months and at certain intervals. It contains the results of 32,917 charges, and 50,360 tons of iron.

Each horizontal line of figures represents monthly returns. Four furnaces are represented, the ore, stone and coke being the same for any one furnace during the period, and all tons of 2,240 lbs.

84   GEOLOGICAL SURVEY OF ALABAMA.

TABLE VI—Illustrative of Coke Furnace Practice With Hard and Soft Red Ore. Increasing Percentage of Hard Ore in Ore Burden of Hard and Soft. Delivery Prices: Hard, 67.5 cts.; Soft, 55.4 cts.; Stone, 63.4 cts.; Coke, $1.75.

Results From the Same Furnace. Consecutive Months. Tons of 2,240 lbs.

Year	Number	Charges	Percent. of Ore Burden		Percent. of Total Burden				Iron Made			Consumption: Tons per ton of iron				Cost per Ton of Iron				Percent. of Cost Per Ton of Iron				Pounds of Coke per Pound of Iron
			Hard	Soft	Hard	Soft	Stone	Coke	Per Charge	Total	Perc't. of Foun-dry Grades	Ore	Stone	Coke	Total	Ore	Stone	Coke	Total	Hard	Soft	Stone	Coke	
1895	16	2954	50.9	49.1	27.7	26.7	15.5	30.1	1.66	4912	99.2	2.39	0.68	1.34	4.41	$1.47	$0.44	$2.31	$4.22	19.5	15.4	10.3	54.8	1.34
1895	17	2370	50.9	49.1	28.1	27.1	15.8	29.0	1.72	4943	98.2	2.51	0.72	1.32	4.55	1.55	0.45	2.31	4.31	19.8	15.7	11.4	53.1	1.32
1895	20	3029	52.3	47.7	27.2	24.8	16.0	32.0	1.62	4882	90.2	2.27	0.69	1.39	4.35	1.48	0.44	2.44	4.36	19.6	14.2	11.3	54.9	1.39
Average.		2951	51.3	48.6	27.7	26.2	15.7	30.7	1.67	4929	95.2	2.39	0.69	1.35	4.43	$1.50	$0.44	$2.35	$4.29	19.6	15.1	11.0	54.3	1.35

THE FUELS.

Results from the same Furnace. Consecutive Months.

1895	19	274	2	51.1	48.9	26	5.25.3	15.8	32	4	1.47	4037	88.6	2.42	0.73	1.52	4.67	$ 1.48	$ 0.45	$ 2.66	$ 4.59	18.1	14.2	11.3	54.9	1.52
1895	18	2708	50.9	49.1	27.0	26.1	15.6	31.3	1.54	4155	68.3	2.54	0.77	1.47	4.78	1.55	0.48	2.58	4.61	18.4	14.0	10.8	55.0	1.47
1895	21	3003	52.3	47.7	26.2	24.1	16.6	33.1	1.49	4495	87.0	2.32	0.73	1.52	4.57	1.43	0.46	2.66	4.55	18.0	13.4	10.4	58.2	1.52
1895	13	2872	48.2	51.8	24.0	26.3	17.2	32.5	1.45	4157	83.9	2.26	0.70	1.57	4.62	1.60	0.49	2.87	4.76	18.7	14.7	10.8	58.0	1.57
Average.		2831	50.6	49.4	25.9	25.4	16.3	32.4	1.49	4211.81.9	2.39	0.75	1.52	4.66	$ 1.51	$ 0.47	$ 2.64	$ 4.62	18.3	14.3	10.8	58.6	1.52

Results from the same Furnace. Consecutive Months.

1890	30	1508	65.9	34.1	19.0	10.3	34.1	1.97	2970	85.7	2.48	0.45	1.52	4.45	$ 1.58	$ 0.28	$ 2.67	$ 4.53	24.4	10.3	8.5	58.8	1.52
1890	28	1343	65.9	34.1	18.7	10.0	34.0	1.95	2615	87.8	2.51	0.46	1.60	4.57	1.58	0.29	2.79	4.66	23.9	10.2	5.3	59.7	1.60
1890	29	1512	65.9	34.1	19.1	9.7	34.3	1.92	2898	83.2	2.57	0.44	1.58	4.59	1.62	0.28	2.75	4.65	24.4	10.4	6.1	59.1	1.58
Average.		1454	65.9	34.1	18.9	10.0	34.5	1.95	2828	92.2	2.52	0.45	1.57	4.54	$ 1.59	$ 0.28	$ 2.74	$ 4.61	24.2	10.3	5.9	59.6	1.57

Results from the same Furnace. Consecutive Months.

1893	38	1895	91.5	8.5	57.3	5.2	2.3	35.2	1.96	3001	83.9	2.78	0.10	1.56	4.44	$ 1.82	$ 0.13	$ 2.70	$ 4.71	36.9	2.8	1.5	58.8	1.56
1893	34	1805	80.7	19.3	47.4	11.2	5.7	35.7	1.83	3315	93.8	2.68	0.28	1.63	4.57	1.93	0.16	2.89	4.99	33.1	5.1	4.0	57.8	1.63
1893	38	1576	100			63.8		36.2	1.91	3005	59.4	2.87		1.03	4.50	1.04		2.85	4.79	40.5			59.5	1.61
Average.		1792	90.7	9.3	56.2	5.5	2.6	35.7	1.90	3407	79.0	2.78	0.12	1.61	4.51	$ 1.89	$ 0.10	$ 2.83	$ 4.82	36.8	2.6	1.8	58.8	1.61

THE FUELS. 87

A critical examination of this table will show:

1st. The amount of ore used per ton of iron made increases with the percentage of hard ore in the burden, rising from 2.39 tons with 61 per cent. to 2.52 tons with 66 per cent, and 2.78 tons with 90 per cent.

2d. The amount of limestone used per ton of iron made decreases with the increase of hard ore, falling from 0.69 ton with 51 per cent., to 0.45 ton with 66 per cent. and 0.12 ton with 90 per cent. With 50 per cent. of hard ore in the ore burden the consumption of stone is 1545 lbs. per ton of iron made, with 66 per cent. of hard ore it is 1008 lbs. and with 90 per cent. of hard ore it is 269 lbs. In one furnace for a period of three months the consumption of stone per ton of iron was 0.75 ton.

3d. The amount of coke used per ton of iron made increases with the increase of lump hard ore, rising from 1.34 tons with 51 per cent. to 1.57 with 66 per cent. and 1.61 with 90 per cent. In the case of one furnace carrying 50.6 per cent. hard the consumption of coke per ton of iron made for a period of three months was 1.52 tons.

Coke is always the most costly ingredient of the burden. In the table under discussion it does not fall below 53 per cent. of the total raw material cost per ton of iron. The tendency towards increasing consumption of coke with increasing amounts of hard ore leads, therefore, to increased costs for raw materials in a ton of iron.

The consumption of coke per ton of iron, the quality of the coke, ore and stone being the same, depends to a very great extent upon the amount of air and its pressure and temperature, which is blown into the furnace per unit of time. Instances are on record in Alabama where the consumption of coke per ton of iron with very heavy lime burdens over considerable periods did not exceed 1.25 tons, but the furnace was well equipped as to boilers, engines and stoves. Under such circumstances it has been said by one of the best furnace men in the Birmingham district that he could use all hard ore (of the best self-fluxing type) and make iron with 1.25 tons of coke without impairing the quality of the iron.

It must, however, be said that the use of crushed hard ore tends to diminish the consumption of coke, for hard ore in large lumps is not easily penetrated by the reducing gases. When a large piece, weighing from 50 to 75 lbs. is exposed to the heat of the furnace in descending the outside of it is first effected. The carbonic acid is removed, the oxide of iron begins to part with its oxygen, and processes of disintegration are set up which continue until the ore is broken into small fragments.

It may be assumed that the oxide of iron is not completely reduced until each piece is exposed to the deoxidizing gases. This takes place with comparative rapidity if the ore is porous, as with certain kinds of brown ore, or if the fragments of ore are sufficiently small. They must not be too small, else the current of gas is checked, the burden packs and the furnace "hangs." But if the size of the ore particles be small enough to allow of easy gas-penetration while not so small as to cause irregularities in the descent of the burden, we should have comparatively favorable conditions for reduction. It would appear that the hard ore has a twofold advantage over the soft ore, first as regards the admixture of lime for making a self-fluxing ore, and second in having the lime combined with carbonic acid. The first advantage renders possible the saving of extraneous lime. Using 80 per cent. of hard ore and 20 per cent. of soft ore in the ore burden there is required 582 lbs. of limestone, as against 1680 lbs. for 50 per cent hard and 50 per cent. soft, a saving of 31 cents per ton of iron in favor of the heavier hard ore burden. This saving, however, may be more than counterbalanced by the greater amount of ore and coke required in the heavier hard ore burden. It may not be possible to obtain better ore, i. e. so far as concerns its iron-content, but it can be improved by crushing. Crushing does not increase the amount of iron but it does increase the reducibility of the ore by enabling the gases from the coke to act upon a larger surface of iron-bearing material. It does more than this. It furthers the evolution of the carbonic acid in the ore, and this renders the ore more porous.

Crushing and calcination have a common purpose, viz: to increase the reducibility of the ore by increasing the amount of iron-bearing surface exposed to the reducing agencies.

The use of crushed hard ore is rapidly extending in Alabama, and it will not be long before the advantages attending its use will force themselves upon those who seem at present to be indifferent to the matter.

In a paper on "Large Furnaces on Alabama material." (Trans. Amer. Inst. Engrs. Vol. XVII. p. 141. 1889) Mr. F. W. Gordon said that the results at Ensley proved the possibility of making a pound of iron with a pound of coke. Since that time and with a better coke than was then used it has happened for a day or so that a pound of coke made a pound of iron, but the coke iron that has been made in Alabama with a ton of coke per ton of iron is insignificent in amount, and there is no reasonable expectation that it will be increased in our day. The present consumption for the best coke is 1.34 lbs. per pound of iron, and this is very near the average between 1.41 and 1.25.

If any hopes were entertained as to the possibility of any one of the Ensley furnace making a pound of iron with a pound of coke even for a week at a time they must long since have been abandoned in the cold light of facts.

4th. The tendency of the percentage of foundry grades of iron is towards a decrease with the increase of hard ore. While this is not strongly accentuated still it appears to be too evident to be neglected. Individual cases may be cited wherein the percentage production of foundry grades during a month was higher when the percentage of hard ore rose to 80 per cent. than when it was at 52 per cent., as by numbers 34 and 20. But on the other hand when the ore burden was composed entirely of hard ore, as in No. 38, the percentrge of foundry grades touched its lowest point, viz. 59.4.

The influence of increasing amounts of hard ore on the quality of the iron is of the utmost importance in the discussion of this subject. Too much stress can not be put on it, for it determines the price at which the product must be sold. The higher the percentage yield of foundry irons the more valuable is the output. Any thing, therefore, that tends to interfere with the make of foundry iron should be most carefully investigated, and conclusions drawn from authentic records must be the chief evidence.

Thirteen cases have been examined, the number of charges being 32,917, and the amount of iron 50,360 tons. Three cases in whioh the percentage of hard ore in the ore burden was 50.9 per cent., 50.9 per cent. and 52.3 show the following percentages of foundry grades respectively, 99.2 per cent., 96.2 per cent., 90.2 per cent., the average being 95.2 per cent.

The total number of charges was 8,853, and the total iron made 14,798 tons.

Four cases in which the percentage of hard ore in the ore burden was 48.2, 50.9, 51.1, and 52.3, show percentages of foundry grades, respectively, 83.9, 68.3, 88.6, and 87.0, the average being 81.9. The number of charges was 11,325, and the iron made 16,845 tons.

In these cases the average percentage of hard ore in the ore burden was 50.6, as against 51.3 in the first case, while the average percentage of foundry grades was 81.9 as against 95.2. While there was a very small difference between these two cases in respect of the amount of hard ore used there was a marked difference in the percentage of foundry grades made, 95.2 per cent. and 81.9 per cent.

Three cases were examined in each of which the percentage of hard ore in the ore burden was 65.9. In one of them with 1,508 charges and 2,970 tons of iron, the percentage of foundry grades was 95.7. In another with 1,343 charges and 2,615 tons of iron the percentage of foundry grades was 87.8. In the third with 1,512 charges and 2,898 tons of iron the percentage of foundry grades was 93.2. The average of 4,363 charges and 8,483 tons of iron was, in foundry grades, 92.2 per cent.

Finally, three cases were examined in which the percentage of hard ore in the ore burden rose from 80.7 to 100. In one of these with 80.7 per cent. hard there were 1,805 charges, 3,315 tons of iron, and 93.8 per cent. of foundry grades. In another with 91.5 per cent. hard there were 1,995 charges, 3,901 tons of iron, and 83.9 per cent. foundry grades. In the third with 100 per cent. of hard there were 1.576 charges, 3,005 tons of iron, and 59.4 per cent. of foundry grades.

Averaging the results from the two furnaces carrying about 50 per cent. of hard ore in the ore burden we find that with 20,178 charges and 31,643 tons of iron the percentage of foundry grades was 88.5.

Comparing this with the results from the furnace carrying 65.9 per cent. of hard ore, with 4,363 charges, 8,483 tons of iron and 92.2 per cent. foundry grades, there seems to be an advantage of 3.7 per cent. foundry grades for the higher percentage of hard ore.

Taking these two together and comparing with them the results from the burden averaging 90 per cent. of hard ore there is found to be a decided falling off in the percentage of foundry grades.

Perhaps all that can now be said is that there seems to be a tendency towards inferior grades of iron when the percentage of hard ore in the ore burden passes 66. The smaller the yield of iron from the furnace the higher is the percentage of foundry grades, and this seems to be independent of the amount of hard ore carried. Out of 8 cases in which the monthly yield was between 3,900 and 5,000 tons there were 37.5 per cent. in which the yield of foundry grades fell below 87 per cent. In 5 cases in which the monthly yield was between 2,500 and 3,400 tons there was only 1, or 20 per cent. in which the percentage of foundry grades fell below 87.

Whether we may conclude from this that rapid driving on a hard ore burden tends to lower grades of iron is not quite clear. Provided that the furnace has sufficient engine power to furnish the requisite blast and stoves enough to furnish the requisite heat there does not seem to be any good reason why she should not work off on foundry grades satisfactorily, even with a very heavy hard ore burden. But to attempt to make high grade iron with hard ore (limey) burdens and insufficient blast, or heat is apt to cause numerous disappointments.

Ore burdens composed of hard, soft and brown ore, the proportion of brown rising from 1.3 per cent. to 100 per cent.

The table embodies the results from 40,270 charges, and 66,653 tons of iron. The delivery prices for the raw materials are as follows, per ton of 2240 lbs.:

Hard ore.................................... 67.5 cents.
Soft ore..................................... 55.4 "
Brown ore ....1.00 "
Coke ....1.75 " .

They are the same as for the table giving the results from ore burdens of hard and soft ore, except that, in addition we have brown ore.

They are not assumed prices but such as were actually paid in the Birmingham District during 1895. Three furnaces are represented, the ore, stone and coke being the same for any one furnace during the period. Each horizontal line of figures represents monthly returns:

TABLE VII—Illustrative of Coke Furnace Practice (Burdens?) with Hard and Soft Red Ore and Brown Ore.

Increasing Percentage of Brown Ore in Ore Burden of Hard, Soft and Brown. Delivery Prices: Hard, 67.5 cts.; Soft, 55 4 cts.; Brown, $1.00; Stone, 63.4 cts.; Coke $1.75. Tons of 2,240 lbs.

Same Furnace. Consecutive Months.

Year.	Number.	Charges.	Percent. of Ore Burden.			Perct. of Total Burden					Iron Made—Tons.		Per cent. of Foundry grades.	Consumption: Tons per Ton of Iron.				Cost per Ton of Iron.				Percent. of Cost per Ton of Iron.				Pounds of Coke per Pound of Iron.
			Hard	Soft	Brown	Hard	Soft	Brown	Stone	Coke	Per Charge	Total		Ore	Stone	Coke	Total	Ore	Stone	Coke	Total	Hard	Soft	Brown	Stone	Coke	
1894	7	2687	33.4	65.3	1.3	17.0	33.1	0.6	16.9	32.4	1.69	4568	85.8	2.22	0.73	1.44	4.39	$1.33	$0.47	$2.50	$4.30	11.7	18.2	0.6	11.5	58.0	1.44
1895	12	2945	47.1	51.5	1.4	24.4	26.7	0.6	16.8	31.5	1.57	4635	97.7	2.37	0.75	1.45	4.57	1.47	0.48	2.54	4.49	16.8	15.0	0.7	10.7	56.8	1.45
1894	15	2690	48.5	50.0	1.5	26.5	27.3	0.7	15.9	29.6	1.79	4818	99.7	2.29	0.64	1.24	4.17	1.41	0.39	2.17	3.97	18.8	15.8	0.8	10.1	54.5	1.24
1895	11	2962	46.5	50.0	3.4	23.2	24.9	1.6	18.7	31.6	1.69	5002	87.6	2.11	0.78	1.34	4.23	1.32	0.49	2.35	4.16	15.9	14.0	1.2	12.6	56.3	1.34
1895	14	2957	48.4	46.8	4.8	25.1	24.3	2.5	16.8	31.5	1.65	4886	99.5	2.30	0.72	1.37	4.32	1.48	0.45	2.37	4.30	12.3	15.0	2.4	15.2	55.1	1.37
1895	10	2833	42.4	50.8	6.8	22.1	26.5	3.7	17.1	30.6	1.64	4659	99.2	2.35	0.77	1.38	4.50	1.49	0.49	2.42	4.40	15.2	15.0	3.6	11.3	54.9	1.38
1895	9	2847	42.1	50.7	7.2	21.8	26.3	3.7	17.6	30.6	1.63	4673	94.5	2.34	0.78	1.38	4.50	1.49	0.51	2.41	4.41	15.0	14.3	3.6	12.5	54.6	1.38
Average.		2833	44.1	52.1	3.8	22.9	27.0	1.9	17.1	32.1	1.67	4745	97.2	2.28	0.74	1.38	4.40	$1.43	$0.47	$2.39	$4.29	15.1	16.3	1.8	12.0	55.3	1.38

THE FUELS. 95

1890-25	1635	57.1	42.5	0.4	31.9	23.8	0.3	10.8	33.4	2.01	3291	83.1	12.48	0.47	48	4.43	$1.54	$0.20	$2.50	$4.42	21.5	13.1	0.2	6.7	58.5	1.48
1890-22	1810	53.7	34.2	12.1	28.5	18.2	6.3	10.6	36.4	1.88	3418	92.0	12.510.49	1.72	4.72	1.69	0.30	3.01	5.00	18.2	9.5	6.0	8.2	00.1	1.72
1890-32	1712	70.3	9.3	20.4	10.0	0.5	2.11.6	6.4	36.8	1.88	3209	86.9	2.55	0.29	61	4.45	1.88	0.18	2.98	4.06	24.6	2.8	10.6	3.7	58.5	1.61
1890-26	1609	58.5	20.4	21.1	32.3	11.2	11.7	8.4	36.4	1.89	3011	90.0	2.5	0.44	52	4.49	1.80	0.22	2.79	4.81	20.3	5.8	10.9	5.9	57.1	1.52
1890-23	1904	53.7	19.2	27.1	29.9	10.7	15.1	9.7	34.6	2.01	3090	89.2	2.40	0.41	49	4.30	1.77	0.20	2.58	4.61	18.8	5.5	14.1	5.9	55.7	1.49
Average.	1754	58.6	25.1	16.3	32.5	13.8	8.0	9.1	35.6	1.93	3386	88.2	2.49	0.42	56	4.47	1.73	0.25	2.77	4.75	20.7	7.3	8.8	6.1	57.5	1.56

Same Furnace. Consecutive Months.

1895-3	1901	17.3	35.1	47.6	8.9	18.2	24.8	19.2	29.1	1.34	2560	92.8	2.37	0.88	32	4.57	$1.86	$0.55	$2.32	$4.73	5.8	9.7	23.8	11.8	48.9	1.32
1894-4	1900	18.3	33.3	48.4	10.0	17.6	25.3	18.6	28.5	1.29	2458	99.2	2.45	0.87	35	4.04	1.01	0.55	2.38	4.84	6.2	9.3	24.3	11.4	48.8	1.36
1895-5	1983	21.0	28.1	50.9	10.5	14.0	25.3	20.3	30.2	1.37	2708	99.2	2.16	0.86	31	4.33	1.73	0.54	2.30	4.57	6.7	7.3	23.9	11.9	50.2	1.31
1895-2	1906	23.8	24.8	51.4	12.4	13.0	26.8	18.2	29.8	1.23	2369	92.1	2.50	0.87	41	4.78	2.04	0.55	2.48	5.07	7.9	6.8	25.4	11.0	48.9	1.41
1895-5	2076	16.0	17.7	66.3	8.8	9.8	36.7	18.1	28.6	1.37	2844	96.1	2.68	0.88	28	4.84	2.33	0.55	2.24	5.12	5.6	5.1	34.7	10.9	43.7	1.29
1895-1	1802			100			52.9	20.4	26.7	1.53	2768	99.1	2.31	0.89	16	4.36	2.31	0.56	2.04	4.91			47.0	11.5	41.5	1.16
Average.	1929	16.1	23.1	60.8	8.4	12.1	31.9	19.1	28.5	1.35	2619	96.9	2.41	0.87	30	4.58	2.03	0.55	2.29	4.87	5.3	6.3	29.8	11.4	47.2	1.30

THE FUELS.   97

A careful examination of the table will sho-.

1st. The amount of brown ore used per ton of iro.. made varies from 2.28 to 2.49 tons. In 1880 the brown ore was not as good as in 1894 and 1895, and the consumption of ore per ton of iron rose to 2.49 tons, although the average percentage of brown ore in the ore burden was 16.3.

With 44.1 per cent. of hard, 52.1 per cent. of soft and 3.8 per cent of brown the consumption of materials per ton of iron was in tons:

Ore..2.28
Stone..0.74
Coke...1.38
                                                    ————
                                                    4.40

and the cost of the materials was:

Ore..$1.43
Stone...0.47
Coke..2.39
                                                    ————
                                                    $4.29

When the proportions were: PER CENT.
Hard...58.6
Soft...25.1
Brown..16.3

the consumption of materials was, in tons per ton of iron:

Ore...2.49
Stone...0.42
Coke..1.56
                                                    ————
                                                    4.47

and the cost per ton of iron was:

Ore...$1.73
Stone...0.25
Coke..2.77
                                                    ————
                                                    $4.75

When the proportions were:
Hard 16.1
Soft 23.1
Brown 60.8
the consumption, in tons per ton of iron, was:
Ore 2.41
Stone 0.87
Coke 1.30
and the cost per ton of iron was:
Ore $2.03
Stone 0.55
Coke 2.29

$4.87

2d. The amount of limestone used per ton of iron varies according to the amount of hard ore used, being 0.42 ton with 58 per cent, 0.74 ton with 44 per cent., and 0.87 ton with 16 per cent. It may be instructive to compare these figures with corresponding results from an ore burden of hard and soft. With 48 per cent. hard in such a burden, which is the nearest to 44 per cent. as above, the consumption of stone in tons per ton of iron was 0.79, as against 0.74 with 44 per cent. of hard in a burden carrying brown ore. The nearest figure in the hard-soft burden to the 58 per cent. hard in the hard-soft brown burden is 65.9 per cent., and this required 0.45 ton of stone per ton of iron, as against 0.42 ton in the brown ore burden carrying 58 per cent. of hard ore.

It is important to note that a hard ore burden with 100 per cent. of hard required no stone, while in the brown ore burden with 100 per cent. of brown the amount of stone required per ton of iron was 0.87 ton, the highest consumption of stone to be observed in these tables.

3d. The amount of coke used per ton of iron decreases with the increase of brown ore, except in the case of the furnace in operation in 1890, and using 58.6 per ce.. hard ore. In this case the consumption of coke was much in excess of the returns for 1894 and 1895, and the general increase of coke with increase of hard ore is borne out also by this table.

4th. The percentage production of foundry iron from brown ore burdens is impaired by increasing the amount of hard ore. With 44 per cent. of hard and 3.8 per cent. of brown ore the average percentage of foundry grades was 97.7. With 58 per cent. hard and 16 per cent. brown it was 88.2 per cent. With 16 per cent. hard and 60 per cent. brown it was 96.9.

As might be expected from the more complex nature of the burden the admixture of hard, soft and brown ores gives rise to greater variations in the economies of production than is the case with burdens of hard and soft ore. The variations are traceable to the fluctuations in the quality of brown ore, for they exhibit wider ranges of composition than either the hard or the soft ore. Then again in physical qualities they are apt to show rapid oscillations. The condition in which brown ore from the same mine and washer reaches the stockhouse has to be observed personally before one can fully appreciate what these may be, and often are. When the brown ore "Bank" is in fairly good ore, and the clay is easily disintegrated, and water is abundant the ore comes in clean. When the clay is "tough," the ore cherty, and the water scanty, the ore comes in wet, and seriously hampered with clay, or with too much insoluble matter.

In spite, however, of these obstacles, which at times may cause trouble, the fact remains that the use of brown ore is highly advantageous. There are very few furnaces that are not glad to get it, and now and then to pay a good deal more than $1.00 per ton for it.

Instances are on record where as much as $1.50 per

ton has been paid in the Birmingham District for brown ore of 55 per cent. iron, although the average price is much lower. Good brown ore always commands a ready sale at fairly remunerative prices.

With the exception of a few furnaces that are not favorably located with respect to hard and soft ore, but are within easy reach of brown ore, the proportion of brown ore used in the coke furnaces rarely exceeds 25 per cent. and for the most part is not above 20 per cent. The ore burden is arranged in various ways, 50 per cent. hard, 25 per cent soft and 25 per cent. brown; 40 per cent hard, 45 per cent soft and 15 per cent. brown; &c. &c.

Under special conditions, such as a large order from pipe-works, &c. the proportion of brown ore is increased until the ore burden may be composed entirely of it. But by far the greater amount of iron made from burdens carrying brown ore is made with about 20 per cent. of brown, hand-picked, and washed, but not calcined.

The practice could be greatly benefited by using washed *and* calcined ore but so far as is known not a single coke furnace is in operation on this kind of material, exclusively or in admixture with hard, and soft ore.

What has been said as to furnace burdens is true in a general way. It is not our purpose now to go into the details of furnace practice, nor to discuss the manner in which the raw materials may be used to the best advantage. This, after all, must be left to the judgment of the furnace manager, which in turn is based on actual experience under varying conditions. It not infrequently happens that one man will take the same materials and the same furnace and produce better iron at a less cost than another, whose theoretical knowledge may be of the best but whose practical acquaintance with the art of making iron has not qualified him to manage a furnace successfully.

There are excellent furnace-men whose knowledge of the difference between silicon and silica is somewhat hazy, and who would find it extremely tiresome to calculate the cubical area of a furnace. They have acquired their information by hard knocks and the exercise of common-sense and a tenacious memory. We have in mind now a good furnace-man who will probably die in the belief that carbonic acid is a combustible material, and who could not calculate the formula of a cinder containing 50 lime, 35 silica and 15 alumina if he was to suffer decapitation the next day.

Iron making is not only a science, it is an art, and one too calling for the constant display of very considerable knowledge and skill, and of untiring patience.

So long as the furnace is working satisfactorily all is well, but to know what to do and when to do it in case something goes wrong, this is what makes or mars the furnace manager.

A furnace may work along weeks at a time on the same burden and produce its normal quantity of iron, and that of a good quality, when some subtle change may take place, discernible only by an experienced eye, and what is to be done must be done at once.

There is one circumstance in connection with iron making in Alabama that renders the daily life of a furnace-man anything but "skittles and beer." It is the wide and at times rapid variation in the quality of the raw materials. The coke is of fairly uniform composition, but the ore is often quite irregular.

There lie before us certain furnace records giving the daily charges of ore, stone and coke over a considerable period. We will take a certain month when the make of iron was 5,719 tons, 77 per cent being foundry grades. There were used 2,503 charges, during the month, a daily average of 80.7.

The furnace was using 80 per cent. of hard ore, and

20 per cent. of soft. During the 31 days the amount of ore in tons per ton of iron varied from 2.62 to 2.19, or 963 lbs. This was during the entire month. From one day to the next there were differences of 600 lbs. of ore per ton of iron. In other words, if the furnace could have been charged every day with ore carrying 45.6% of iron, as was the case on one day, the yield of iron in the month could have been 6,620 tons instead of 5,719, a difference in favor of the better ore of 901 tons for the month. The daily production of iron could have been 213 tons instead of 184 tons.

Furthermore. Not only is the daily yield of the furnace seriously hampered by such irregularities in the ore, the percentage of foundry iron in the make is also lessened, and there are opportunities for an increased consumption of coke and greater costs of production.

In burdening a furnace it is in every way better to to have a leaner ore of regular composition than a richer ore of variable and varying composition.

There would be fewer and more restricted variations in the cost accounts, and less interference with the production of the better grades of iron in the one case than in the other.

The question of securing ore of more contant composition is one that can not be brought too forcibly to the attention of iron makers in Alabama. It dominates all other considerations, and is to-day the most vital problem confronting them. No other single question is at once so important and so little studied, the interest in it seeming to be in inverse proportion to its gravity.

Charcoal Furnace Burdens.

The production of charcoal iron is diminishing in this State, partly because of the increasing proportion of coke iron going to car-wheels and such products, and partly because of the increasing cost of charcoal. The reputation of the charcoal iron mode in the State has been most excellent, especially that of Shelby furnaces, and even now in these times of depression the Shelby iron is sought for by those who still desire a high grade charcoal iron.

The charcoal used is made for the most part in the old way, in mounds and heaps, the attempt to recover by products in specially constructed kilns being confined to the Round Mountain Company in Cherokee county.

By far the greater amount of charcoal iron is derived from the brown ores, the consumption of ore per ton of iron being from 1.80 to 2.03 tons.

The following table exhibits the furnace burdens in good practice over a period of 4 months, with brown ore :

TABLE VIII—CHARCOAL FURNACE PRACTICE.

YEAR	Number	Percentage of total burden.			Iron Made Total.	Consumption: bushels for coal. per ton of iron;			Cost per Ton of Iron.				Pounds of Coal per Pound of Iron.	Per cent. of yield of iron Nos. 1 to 5 inclusive.	Per cent. of Cost Per Ton of Iron.		
		Ore.	Stone.	Coal.		Ore.	Stone.	Coal.	Ore.	Stone.	Coal.	Total.			Ore.	Stone.	Coal.
1893	46	58.2	9.8	32.0	1712	1.84	0.31	100.8	$1.78	$0.26	$6.55	$8.59	0.99	78.0	20.7	2.7	76.6
1894	47	60.0	10.6	29.4	1753	1.95	0.34	97.2	1.72	0.26	5.73	7.71	0.96	90.4	22.3	3.4	74.3
1895	48	59.6	10.9	29.5	1789	2.03	0.37	101.8	1.85	0.28	5.85	7.78	1.00	97.7	21.2	3.6	75.2
1896	49	58.3	9.6	32.1	1893	1.80	0.30	100.7	1.50	0.24	5.63	7.37	0.99	88.9	20.3	3.3	76.4
Avg.		59.0	10.2	30.8	1787	1.90	0.33	100.1	1.66	0.26	5.94	7.86	0.99	88.7	21.1	3.3	75.6

In this table 2,748 cubic inches are taken as one bushel, and the weight per bushel is taken at 22.4 lbs., so that 100 bushels equal one ton of 2,240 lbs.

According to these returns the average percentage, furnace yield, of iron in these brown ores was 52.6, the average consumption of ore per ton of iron being 1.90 ton; the average consumption of limestone was 0.33 ton, or 739 pounds; and of charcoal 100.1 bushels.

The ore was partly washed and calcined, partly merely washed, No. 46 being washed and calcined.

Investigations that have been carried on for some months, but which are not yet to be published, have shown that there is a marked decrease in the amount of charcoal required per ton of iron and a decided increase in the output of the furnace consequent upon the use of washed and calcined ore. This may not appear from the examination of the returns of a single month, as for instance in No. 46. But after comparing the same ore under these different conditions, the other elements of practice being the same, there is no room for doubt.

The charcoal furnaces have the advantage over the coke furnaces of much better ore, but their fuel is far more costly than coke, and the percentage cost of the fuel is considerable more than with coke iron.

Charcoal iron is worth more than coke iron, the present selling price being about twice as much for the one as for the other. The entire product is consumed by manufacturers of car wheels, and those who make a specialty of tough, chilled castings. In the old days a great deal of charcoal iron was used in boiler plates, but the increasing use of soft steel for this purpose has gradually destroyed this business, and very little of it now goes to boiler works.

CHAPTER VI.

FURNACES, ROLLING MILLS, &c.

Coke Furnaces in Alabama.

(From the Directory of the Iron and Steel Works in the United States, Amer. Iron and Steel Assoc., Phila., 1896. Jas. M. Swank, M'g'r.)

Bessemer Land and Improvement Company, Bessemer, Jefferson county. One completed stack, one partly erected, and two projected. Fort Payne Furnace, Fort Payne, KeKalb county, one stack, 65 x 14, built in 1889-90, and blown in September 3d, 1890, three Siemens-Cowper-Cochrane stoves; fuel, coke; ores, red and brown hematite; product, forge and foundry pig iron; annual capacity, 27,000 gross tons; (formerly operated by the Fort Payne Furnace Company).

Bay State Furnace, Fort Payne, DeKalb county, one stack, 65x14, partly erected in 1890-1 by the Bay State Furnace Company; work suspended in 1891; three firebrick stoves; furnace may be completed by the present owner, or it may be torn down and removed to Bessemer. Company also contemplates erecting immediately, at Bessemer, two stacks, each 75x17, and several hundred coke ovens.

Edwards Furnace, at Woodstock, Bibb county, one stack 70x15, first blown in June 10, 1880; remodeled in 1887 and 1890; three hot blast stoves; stack to be torn down and removed to Bessemer; formerly operated by the Edwards Iron Company. H. F. DeBardeleben, President; Walker Percy, Vice-President; H. M. McNutt, Secretary and Treasurer.

Clara Furnace, F. W. Roebling, Trustee for Bondholders, Trenton, N. J. Furnace at Birmingham, Jefferson county. One stack 65x15.5; commenced building February 9th, 1890; blown in August 23d, 1890; three Massicks and Crooke stoves; fuel, coke; ores, brown and soft and hard red from Alabama and Georgia; product, a strong low-phosphorous foundry pig iron; annual capacity, 22,500 gross tons. For sale.

Clifton Furnaces, Clifton Iron Company, Ironaton, Talladega county. Two stacks: No. 1, 55x13, changing to 70x16, built in 1884, blown in April 16, 1885; No. 2, 60x14, built in 1889–90,and blown in during 1891; built to use charcoal for fuel, but changed to coke in 1895; six Cowper stoves; fuel, Alabama coke; ore, local brown hematite; product, foundry pig iron; total annual capacity, 72,000 gross tons. Brand, "Clifton." T. G. Bush, President, Anniston; Augustus Lowell, Vice-President, Boston, Mass.; Paul Roberts, Secretary and Assistant Treasurer, Ironaton. Selling agents, Matthew Addy and Co., Cincinnati; C. L. Pierson and Co., Boston and New York.

Gadsden-Alabama Furnace, Gadsden, Etowah county. One stack, 75x16, built in 1887–88, and first blown in October 14, 1888; three Whitwell stoves; fuel, coke; ores, local red and brown hematite; product, foundry and basic pig iron; annual capacity, 35,000 gross tons. Brand, "Etowah." Owned by Thomas T. Hillman, George L. Morris and Mrs. Aileen Ligon, of Birmingham. Idle, and for sale or lease.

Hattie Ensley Furnace, Colbert Iron Company, lessee, Sheffield, Colbert county. One stack, 75 x 17, built in 1887, and blown in December 31st, 1887; three Whitwell stoves; fuel, coke; ore, local brown hematite; product, foundry pig iron, annual capacity 48,000 gross tons.. Brand, "Lady Ensley." A. A. Berger, President; Wade Allen, Vice-President; J. V. Allen, Secretary and Treasurer; A. J. McGarry, Manager. Selling agents, Lee Chamberlain and Co., Columbus, Ohio. Owned by the James P. Witherow Company, Pittsburg.

Lady Ensley Furnace, Lady Ensley Furnace Company, Sheffield, Colbert county. One stack, 65 x 17, built in 1887-90, and first blown in April 25th, 1889 ; three Whitwell stoves; annual capacity 45,000 gross tons. R. W. Cobb, Receiver. Idle since June, 1892.

Mary Pratt Furnace, W. T. Underwood, Birmingham, Jefferson county. One stack, 65 x 14, built in 1882, and first put in blast in April, 1883; rebuilt in 1889; three Whitwell stoves; fuel, coke; ores, local brown and red fossiliferous ; annual capacity, 30,000 gross tons. Brand, "Mary Pratt." Idle.

Philadelphia Furnace, Florence Cotton and Iron Company, Florence, Lauderdale county. Main office, 330 Walnut St., Phila. One stack, 75 x 17, commenced by the W. B. Wood Furnace Company in 1887, and completed by the present company in 1890-91; three Whitwell stoves, each 70 x 20 ; fuel, coke ; ore, brown hematite from Lawrence county, Tenn.; product, foundry pig iron ; annual capacity 45,000 gross tons. Brand, "Philadelphia." Abraham S. Patterson, President, Robert Dornan, Vice-President; James Pollock, Secretary and Treasurer. For sale.

Pioneer Furnaces, Pioneer Mining and Manufacturing Company, Thomas, Jefferson county. Two stacks, each 75 x 16.5; No. 1 built in 1886-88, and blown in May 15, 1888 ; No. 2 built in 1889-90, and blown in February 22nd, 1890 ; eight Siemens-Cowper-Cochrane stoves; fuel, Alabama coke; ores, red and brown hematite from the company's mines near the furnaces; product, foundry pig iron; total annual capacity, 95,000 gross tons. Brand, "Pioneer." Edwin Thomas, President, and Samuel Thomas, Vice-President, Catasaqua, Penna.; George H. Myers, Secretary and Treasurer, Bethlehem,

Penna. Selling agents, Matthew Addy and Co., 201 Walnut Place, Phila.

Sheffield Furnaces, Sheffield Coal, Iron and Steel Company, Sheffield, Colbert county. Three stacks, each 75 x 18, built in 1887-88; No. 1 blown in during Sept., 1888, and No. 2 blown in during Oct., 1889; No. 3 not yet blown in; Nos. 1 and 2 rebuilt in 1891; nine Whitwell-Cowper stoves; fuel, Alabama and Virginia coke; ores, Alabama and Tennessee brown hematite; product, foundry pig iron; total annual capacity, 150,000 gross tons. Brand, "Sheffield." E. W. Cole, President, Nashville, Tenn.; Jerome Keeley, Vice-President, Phila., George H. Berlin, Secretary, Sheffield; J. A. McKee, Treasurer, Phila.; Samuel Adams, Superintendent, Sheffield. Selling agents, Rogers, Brown and Co., Cincinnati.

Sloss Furnace, Sloss Iron and Steel Company, Birmingham, Jefferson county. Four stacks: No. 1, 82.25x18, built in 1881 82, put in blast April 12th, 1882, and rebuilt in 1895; No. 2, 68x18, built in 1882; No. 3, 73x16.5, built in 1887-88, and blown in during Oct., 1888; No. 4, 73x16.5, built in 1887–89, and blown in during Feb., 1889; five Whitwell, eight Gordon-Whitwell-Cowper, and three two-pass 18x70 stoves; fuel, coke; ores, red fossiliferous, hard and soft, and brown hematite; ores and coal mined on the company's property within ten to fifteen miles of furnaces; product, foundry and mill pig iron; total annual capacity, 200,000 gross tons. Brand, "Sloss." Thomas Seddon, President; E. W. Rucker, Vice-President; W. L. Sims, Secretary and Treasurer; J. H. McCune, Furnace Manager. Selling agents, D. L. Cobb, Louisville and Chicago; J. E. Cartwright, St. Louis; Rogers, Brown and Warner, Phila.; Hugh W. Adams and Co., 15 Beekman St., N. Y.

Spathite Furnace, The Spathite Iron Company, Florence, Lauderdale county. One stack, 75x14, completed in December, 1888, and blown in during Oct., 1889; rebuilt in 1893; three improved Pollock stoves; fuel, coke; ores, spathite and brown hematite from Iron City, Tenn.; product, spathite pig iron; annual capacity, 30,000 gross tons. Brand, "Spathite." (Formerly called North Alabama Furnace.) J. Overton Ewin, Receiver; J. H. Short, Superintendent. Selling agents, Rogers, Brown & Co., Cincinnati. Sold Nov. 25th, 1895, to Louisville Banking Company, Louisville, Kentucky.

Talladega Furnace, Talladega Furnace Company, Talladega, Talladega county. One stack, 72x18, built in 1889, and blown in October 5th, 1889; three Ford and Moncur stoves, each 62x26; fuel, Alabama and West Virginia coke; ore, local brown hematite; product, Bessemer, foundry and forge pig iron; annual capacity, 40,000 gross tons. Brand, "Talladega." W. P. Armstrong, President; George Dunglinson, Secretary; R. L. Ivey, Treasurer. Negotiations now pending for the sale of the furnace.

Tennessee Coal, Iron and Railroad Company, Birmingham, Jefferson county. Thirteen stacks in Jefferson county. Five stacks at Bessemer: Nos. 1 and 2, each 75x17, built in 1886-87; No. 1 put in blast in 1888, and No. 2 in 1889; seven Whitwell stoves; Nos. 3 and 4, each 75x17, built in 1889–90; eight Whitwell stoves; No. 5, or Little Belle, 60x12, built in 1889–90, three Whitwell stoves.

Oxmoor Furnaces, at Oxmoor, (formerly called Eureka Furnaces) two stacks: No. 1, 75x17, completed in July, 1877, and rebuilt and blown in during Dec., 1885; No. 2, 75x17, first blown in in March, 1876, and rebuilt and blown in during Aug, 1886; seven Whitwell stoves. Fuel, Alabama coke, made in the company's ovens; ores, local brown hematite and red fossiliferous from the company's mines; product, foundry, mill and basic open-hearth pig iron ; total annual capacity, 361,500 gross tons. Brand, "DeBardeleben."

Alice Furnaces, at Birmingham, two stacks : No. 1, 75x15, built in 1879-80,-and put in blast November 23d, 1880; raised to present height in 1890; three Gordon-Whitwell-Cowper stoves; No. 2, 80x17.5, built in 1883, and put in blast July 24th, 1883; three Whitwell stoves; brand, "Alice."

Ensley Furnaces, at Ensley. Four stacks, each 80x19.5, built in 1887, 1888, and 1889 ; No. 1 blown in March 19, 1889 ; No. 2, December 1st, 1888; No. 3, June 5th, 1888, and No. 4 April 9th, 1888; four Gordon-Whitwell-Cowper stoves to each furnace. Brand, "Ensley." Fuel, coke made in the company's ovens; ores, red and brown hematite from the company's mines at Hillman, Redding and Woodstock ; product, foundry, mill and basic open-hearth pig iron ; annual capacity of Alice Furnaces 113,000 gross tons ; of Ensley Furnaces, 270,000 tons. Total annual capacity of the thirteen stacks, 744,000 tons. N. Baxter, Jr., President; David Roberts, 1st Vice-President; A. M. Shook, 2d Vice-President ; George B. McCormack, General Manager; James Brown, Treasurer ; Andrew M. Adger, Secretary and Assistant Treasurer ; H. D. Cooper, Auditor; Erskine Ramsay, Chief Engineer; John Dowling, Superintendent of Bessemer Division; A. E. Barton, Superintendent of Ensley Division. Selling agents, Rogers, Brown & Co., Cincinnati, and branch houses; Matthew Addy & Co., Cincinnati and St. Louis.

Trussville Furnace, Trussville, Jefferson county. One stack, 65x18, built in 1887-89, and blown in in April, 1889; three Whitwell stoves ; fuel, Alabama coke ; ore, local red hematite ; product, foundry pig iron; annual capacity, 30,000 gross tons. Brand, "Trussville." Owned by Messrs. Hogsett, Ewing and Thompson. Negotiations pending for the sale of the furnace ; if sold, the stack will be enlarged to 75x18, and the capacity increased to 60,000 gross tons.

Williamson Furnace, Williamson Iron Company, Birmingham, Jefferson county. One stack, 65x13.66, built in 1886, and first blown in in October, 1886; three Massicks and Crooke stoves; fuel, coke made at Coalburg; ores, red fossil and brown hematite; product, foundry and mill pig iron; annual capacity, 18,000 gross tons. Brand, "Williamson." C. P. Williamson, President and General Manager; H. D. Williamson, Vice-President; J. B. Simpson, Secretary and Treasurer.

Woodstock Furnaces, The Woodstock Iron Works, Anniston, Calhoun county. Two stacks, each 75x16, built in 1887–89, and one blown in October 10th, 1889; six Whitwell stoves; fuel, Alabama coke; ore, local brown hematite; product, foundry pig iron; total annual capacity, 72,000 gross tons. Brand "Woodstock." John D. Probst, President, and George Glover, Secretary, New York; H. Atkinson, Vice-President; J. W. McCulloh, General Manager, and W. L. Doane, Treasurer and Assistant Secretary, Anniston. Altering and repairing one stack, to increase daily capacity from 125 to 165 tons.

Woodward Iron Company, Woodward, Jefferson county. Two stacks, each 75x17, one built in 1882–83, and put in blast in August, 1883, and the other built in 1886; eight Whitwell stoves; fuel, coke made from the company's coal; ore, red fossiliferous, mined within three miles of the furnace; specialty, foundry pig iron; total annual capacity, 100,000 gross tons. Brand, "Woodward." J. H. Woodward, President; Frank M. Eaton, Secretary; Silas Hine, Treasurer.

Number of coke furnaces in Alabama, 39 completed stacks, 1 stack partly erected, and 2 stacks projected.

Annual capacity of coke furnaces in Alabama, 1,804,000 gross tons.

Number of coke and bituminous furnaces in the

United States, 256; annual capacity, 13,118,600 gross tons.

Alabama has 15.2 per cent. of the total number of coke furnaces, 13.7 per cent. of the total annual capacity, and produces 10.5 per cent of the total amount of coke iron.

Dividing the period 1876–1895 into 4 sub-periods of 5 years each we have the following comparisons:

1876–1880, coke furnaces built 4, production in 1876 1,262 tons, in 1880 35,232 tons, increase 33,970 tons, or 28 times.

1881–1885, coke furnaces built 6, production in 1881 48,107 tons, in 1885 133,808 tons, increase 85,701 tons, or 2.78 times.

1886–1890, coke furnaces built 29, production in 1886 180,133 tons, in 1890 718,383 tons, increase 538,250 tons, or 3.99 times.

1891–1895, no coke furnaces built.

The greatest activity was displayed in the period 1886 1890, as of the 39 completed stacks in 1895 29, or 74.5 per cent. were built during these years. It was not until 1888 that the production of coke iron passed the 200,000 ton mark, and not until 1889 did it rise above 500,000 tons, and assume respectable proportions. The year 1895 witnessed the largest production of coke iron ever recorded in the State, 835,851 tons, excelling the output of 1892 by 11 tons.

Of the 835,851 tons 387,793 tons, 46.4 per cent. were made during the first half of the year, 18 furnaces being in blast June 30th, and 448,058 tons 53.6 per cent. in the second half, 20 furnaces being in blast December 31st.

The 60,265 tons made in the second half of the year in excess of the output during the first half may be taken as representing the increase due to the upward tendency of prices which seemed to be genuine about that time.

The production of coke iron since 1876 is given in the following table:

TABLE IX.
Production of Coke Iron in Alabama.—Tons of 2,240 pounds.

Year.	Tons.	Year.	Tons.	Year.	Tons	Year.	Tons.
1876	1,262	1881	48,107	1886	180,133	1891	717,687
1877	14,643	1882	51,093	1887	176,374	1892	835,840
1878	15,615	1883	102,750	1888	317,289	1893	659,725
1879	15,937	1884	116,264	1889	608,034	1894	556,314
1880	35,232	1885	133,808	1890	718,383	1895	835,851

Charcoal Furnaces in Alabama.

[From the Directory to the Iron and Steel Works in the United States, American Iron and Steel Association, Phila. 1896. Jas. M. Swank, Manager.]

Attalla Furnace, Buffalo Iron Company, Nashville, Tenn. Furnace at Attalla, Etowah County. One stack, 55x11, built in 1888-89, and blown in June 15th, 1889; iron stoves: ores, red and brown hematite from Etowah and Cherokee counties; product, car-wheel pig iron; annual capacity, 18,000 gross tons. Brand "Attalla". Robt. Ewing, President; J. A. Cooper, Secretary and Treasurer.

Bibb Furnace, Alabama Iron and Steel Company, Brierfield, Bibb county. One stack, 55x12, built in 1864 to use charcoal; re-built in 1881, and remodeled in

1886 to use coke; returned to the use of charcoal in 1890; re-built in 1892; warm blast; ore, brown hematite, mined in the vicinity; product, car-wheel pig iron; annual capacity, 14,500 gross tons. Brand, "Bibb". T. J. Peter, President. Selling agents, C. R. Baird & Co. Phil., De Camp & Yule, St. Louis; Forster, Hawes & Co., Chicago.

Coosa Furnace, Gadsden Iron Company, Gadsden, Etowah county. One stack, 64x12, built in 1882 with material from the Vigo Iron Company's No.1 furnace at Terra Haute, Indiana; first blown in May 30th, 1883; hot blast; ores, local red and brown hematite; product, foundry and car-wheel pig iron; annual capacity, 8,000 gross tons. Brand, "Stewart". (Formerly called Gadsden Furnace). A. J. Crawford, President, Terre Haute, Ind.; T. W. Stewart, Secretary, Treasurer, and General Manager.

Decatur Charcoal Iron Furnace, The Decatur Land Company, New Decatur, Morgan county. One stack, 60x12, built in 1887-88, and blown in February 23d, 1890; two Gordon-Whitwell-Cowper stoves; used coke as fuel for a short time; ores, red and brown hematite; annual capacity, 18,000 gross tons. J. D. Probst, President, and C. Y. Kent, Assistant Secretary, Nsw York; C. C. Harris, Vice-Prssident: W. T. Mulligan, Secretary, W. A. Bibb, Treasurer, John C. Eyster, General Counsel, New Decatur. For sale, or lease.

Jenifer Furnace, Jenifer Furnace Company, Jenifer, Talladega county. Central office, Anniston. One stack, 56x11, built in 1892, and blown in December 5th 1892, taking the place of the old stone stack built in 1863; two Hugh Kennedy stoves; each 45x16; ore, local brown hematite; product, car-wheel pig iron; annual capacity 12,000 gross tons. Brand "Jenifer". (One stack, built in 1863, abandoned and dismantled in 1872).

John H. Noble, President, and John E. Ware, Secretary and Treasurer, Anniston. Selling agents, Rogers, Brown & Co., Cincinnati, and St. Louis; C. R. Baird & Co., Phila.

Langdon Furnace, (once known as Stonewall Furnace.) The National Bank of Augusta, Augusta, Ga. Furnace at Langdon, (P. O. address, Rock Run Station), Cherokee county. One stack, 42x11, built in 1873, and re-built in 1889–90; blown in in May 1890; one stove; ore, local brown hematite; product, car-wheel pig iron; annual capacity, 12,000 gross tons. Brand, "Langdon". For sale.

Piedmont Land and Improvement Company, Piedmont, Calhoun county. Commenced in 1890 the erection of one stack, 60x12, with two Gordon-Whitwell-Cowper stoves; work suspended in 1891; expects to complete stack in 1896. W. P. Smalley, President; J. H. Ledbetter, Vice-President; R. L. Hurt, Secretary and Treasurer.

Rock Run Furnace, Rock Run Iron and Mining Company, Rock Run, Cherokee county. One stack, 54.5x11.5, built in 1873–4, enlarged in 1881 and 1892, and rebuilt in 1894; warm blast; ore, local brown hematite; product, car-wheel pig iron; annual capacity, 15,000 gross tons. Brand, "Rock Run." J. H. Bass, President, J. I. White, Secretary, and F. S. Lightfoot, Treasurer, Fort Wayne, Indiana; J. M. Garvin, Superintendent, Rock Run.

Round Mountain Furnace, (Formerly called Round Mountain Iron Works,) The Round Mountain Furnace Company, lessee, Chattanooga, Tenn. Furnace at Round Mountain, Cherokee county. One stack, 45x9.5, built in 1853, rebuilt in 1874, and remodeled in 1888; cold blast; ore red fossilliferous; specialty, cold blast charcoal pig iron for chilled rolls and car-wheels; annual capaicty, 6,500 gross tons. Brand, "Round Mountain." L. S. Colyar, President; Jo. C. Guild, Vice-President: E. Shackelford, Secretary; E. B. Pennington, Superintendent. Selling agents, Rogers, Brown & Co., Cincinnati and branch houses; J. E. Cartwright, St. Louis. Owned by the Elliott Pig Iron Company, Gadsden.

Shelby Furnaces, Shelby Iron Company, Shelby, Shelby county. Two stacks, Nos. 1 and 2, each 60x14, built in 1863 and 1873; No. 1 rebuilt in 1889; warm blast; ore, brown hematite obtained on the furnace property; product, car-wheel pig iron; total annual capacity, 40,000 gross tons. Brand, "Shelby." T. G. Bush, President, Anniston; B. Y. Frost, Secretary, and W. S. Gurnee, Treasurer, 80 Broadway, N..Y.; E. T. Witherby, Assistant Treasurer, Shelby. Selling agents, Matthew Addy & Co., Cincinnati; C. L. Pierson & Co., Boston and New York.

Tecumseh Furnace, Tecumseh Iron Company, Tecumseh, Cherokee county. One stack, 60x12, built in 1873, and blown in February 19th, 1874; hot blast; ore, local brown hematite; annual capacity, 13,500 gross tons. Brand, "Tecumseh." P. N. Moore, President; S. J. Fearing, Treasurer and General Manager. Idle since October, 1890; for lease.

Woodstock Furnace, Woodstock Iron Works, Anniston, Calhoun county. One stack, 50x12, blown in April 13th, 1893; rebuilt in 1880; hot blast; ore, local brown hematite; product, car-wheel pig iron; annual capacity, 11,000 gross tons. Brand, "Woodstock." (One stack partly destroyed by fire in 1891.) John D. Probst, President, and George Glover, Secretary, N. Y.; H. Atkinson, Vice-President, J. W. McCullough, General Manager, W. L. Doane, Treasurer and Assistant Secretary, Anniston. May dismantle both stacks.

Number of charcoal furnaces in Alabama, 12 completed stacks, and 1 stack partly erected, annual capacity, 168,500 gross tons. Number of charcoal furnaces in the United States, 96 ; annual capacity, 1,098,550 gross tons. Dividing the period 1876–1895, as under coke furnaces, into 4 sub-periods of 5 years each, we have the following comparisons :

1876–1880—charcoal furnaces built, 1; output in 1876, 20,818 tons ; in 1880, 33,693 tons ; increase, 12,875, or 1.62 times.

1881–1885—charcoal furnaces built, 2 ; output in 1881, 39,483 tons ; in 1885, 69,261 tons ; increase, 29,778 tons, or 1.75 times.

1886–1890—charcoal furnaces built, 4 ; output in 1886, 73,312 tons, in 1890, 98,528 tons ; increase, 25,216 tons, or 1.34 times.

1891–1895—charcoal furnaces built, 2 ; output in 1891, 77,985 tons ; in 1895, 18,816 tons ; decrease 59,169 tons, or a little more than three-fourths.

Of the twelve completed charcoal stacks in 1895, 4, or 33 per cent. were built in the period 1886–1890, two in 1873, one in 1874, and the others as above. In charcoal, as in coke furnaces, the greatest activity was displayed during the period 1886–1890, although the activity in coke furnaces was much more pronounced. Alabama has 72.5 per cent. of the total number of charcoal furnaces, 15.3 per cent. of the total annual capacity, and made in 1895 8.3 per cent. of the total production of charcoal iron.

The charcoal iron industry has been declining for several years. It reached its maximum in 1889, with 98,595 tons. At that time Alabama was producing 17.1 per cent. of the total, and was second in point of production.

The statistics of production are given in the following table :

FURNACES, ROLLING MILLS, &C.   119

TABLE X.
Product of Charcoal Iron in Alabama. Tons of 2,240 Pounds.

Year.	Tons.	Year.	Tons.	Year.	Tons.	Year.	Tons.
1872	11.171	1878	21.422	1884	53.078	1890	98.528
1873	19.895	1879	28.563	1885	69.261	1891	77.985
1874	29.342	1880	33.763	1886	73.312	1892	79.456
1875	22.418	1881	39.483	1887	85.020	1893	67.163
1876	20.818	1882	49.590	1888	84.041	1894	36.078
1877	22.180	1883	51.237	1889	98.595	1895	18.816

Hot Blast Stoves in Alabama—1896.

Cowper.		Ford and Moncur.		Gordon-Whitwell-Cowper.		Hugh Kennedy.		Massicks and Crooke.		Pollock.		Seimens-Cowper-Cockrane.		Whitwell.		Whitwell-Cowper.		Total
No.	Per Cent.	No.	Per Cent.	No.	Per Cent.	No.	Per Cent.	No.	Per Cent.	No.	Per Cent.	No.	Per Cent.	No.	Per Cent.	No.	Per Cent.	
6	4.4	3	2.2	31	22.8	2	1.5	6	4.4	3	2.2	11	8.1	65	47.8	9	6.6	136

Rolling Mills, Steel Works, &c., in Alabama.

(From the Directory of the Iron and Steel Works in the United States. American Iron and Steel Assoc., Phila., 1896, Jas. M. Swank, Manager.)

Alabama Iron and Steel Company (Formerly Brierfield Rolling Mill,) Brierfield, Bibb county. Built in 1863, rebuilt in 1882–83, and put in operation in August, 1883; 10 double and 4 single puddling furnaces, 5 heating furnaces, 3 18-inch trains of rolls, and 72 cut nail machines; product, merchant bar iron and nails; annual capacity, 12,000 gross tons. T. J. Peter, President.

Alabama Rolling Mill Company, Birmingham, Jefferson County. Works at Gate City, Jefferson county. Built in 1887–88 and put in operation in February, 1888; 23 single puddling furnaces, 2 gas heating furnaces, and 3 trains of rolls (18-inch muck and 8 and 16 inch bar); product, bars, bands, hoops, light T rails, &c.; annual capacity, 24,000 gross tons. W. J. Behan, President; W. H. Hassinger, Vice-President and General Manager; D. M. Forker, Secretary and Treasurer.

Alabama Steel Works, (formerly Fort Payne Rolling Mill,) The DeKalb Company, lessee, Fort Payne, DeKalb county. Built in 1889-90; two 15-grosst on basic open-hearth steel furnaces; first steel made in July, 1893; 4 gas heating furnaces, 5 cut-nail machines, (idle,) and 2 trains of rolls (one 2-high 32-inch reversing and one 22-inch nail plate); product ingots, blooms, billets and slabs; annual capacity, 10,000 gross tons of ingots. Fuel used, producer gas. E. N. Cullom, President; H. A. Yeaton, Treasurer; S. C. Adams Secretary. Owned by the Alabama Steel Works, (incorporated).

Annisto Rolling Mills, Anniston Iron and Steel Company, lessee, Anniston, Calhoun county. Built in 1890-91; 12 single puddling furnaces, 2 large heating furnaces and 2 trains of rolls, (3-high 20-inch muck and 3-high 12-inch finishing) J. K. Dimmick, President; H. B. Cooper, Vice-President and General Manager; John S. Mooring, Secretary and Treasurer. Owned by the Anniston Rolling Mills Company.

Bessemer (The) Rolling Mills, Bessemer, Jefferson county. Built in 1887-88; 24 single puddling furnaces, 6 heating furnaces, 5 trains of rolls (one 20-inch muck, one 8-inch guide, one 16-inch car, one 22-inch sheet, and one 26-inch plate), and 3 Siemens gas producers; product, bar, guide, plate and sheet iron; annual capacity, 27,000 gross tons. Owned by Morris Adler, of Birmingham, and others. Idle since the spring of 1891, and for sale.

Birmingham Rolling Mills, Birmingham Rolling Mill Company, Birmingham, Jefferson county. Built in

1880, and first put into operation in July, 1880; enlarged in 1887 and 1895; 11 double and 24 single puddling furnaces, one scrap furnace, 7 gas, 4 box annealing, 2 pair, and 4 sheet heating and annealing furnaces, and 9 trains of rolls, (two 8-inch guide, one 16-inch bar, two 18-inch forge, two 24-inch sheet, one 26-inch plate, and one 24-inch finishing); product, iron and steel bars, plates, sheets, angles, round-edge tire, small T rails, fish plates, &c.; annual capacity, 70,000 gross tons. Fuel used, producer gas and coal. Contemplates erecting an open hearth steel plant (basic). James G. Caldwell, President; Thomas Ward, General Manager; J. D. Dwyer, Superintendent; J. H. Mohns, Salesman.

Jefferson Steel Company, Birmingham, Jefferson county. Built in 1889–90; one 15-gross ton basic open-hearth steel furnace; first steel made April 14th, 1890; product, ingots; annual capacity, 8,100 gross tons. Brand "Jefferson." (This furnace takes the place of one experimental Henderson open-hearth steel furnace built in 1887–88, and first steel made February 27th, 1888. Formerly operated by the Henderson Steel and Manufacturing Company.) Eugene F. Enslen, President; P. A. Buyck, Vice-President; McK. Thomas, Secretary, Treasurer and General Manager.

Shelby Rolling Mill Company (formerly Central Iron Works), Helena, Shelby county. Works started in March, 1873; enlarged by present company in 1889; 10 single puddling furnaces, 3 heating furnaces, and 4 trains of rolls; product, merchant bar and band iron, and light T rails; annual capacity, 7,200 gross tons. Company failed; works idle for several years. Address, Joseph F. Johnston, Birmingham.

United States (The) Car Company, Anniston, Calhoun county. Chicago office, 1480 Old Colony

Building; New York office, 45 Broadway. Built in 1884 and enlarged in 1888-89, and 1893; one single and six double puddling furnaces, six heating furnaces, one scrap furnace, two trains of rolls (one 18-inch muck and bar, and one 10-inch merchant and guide), and five hammers (one 6,000 lb., two 4,000 lb., and two helve); product, car axles and merchant bar iron; annual capacity, 15,000 gross tons. David Cornfoot, President, London, England; Thomas Sturgis, Vice-President, New York; J. M. Maris, General Manager, Chicago; O. M. Stinson, General Superintendent, Anniston.

Steel Works Projected.

Bessemer Land and Improvement Company, Bessemer, Jefferson county, contemplates erecting an open-hearth (basic) steel plant at Bessemer in the spring or summer of 1896.

The Tennessee Coal, Iron & Railroad Company, in connection with the Louisville & Nashville Railroad Company, the Southern Railway Company and private persons in Birmingham, contemplates the erection of a large basic open-hearth steel plant during 1896-97. Location not yet chosen.

Number of rolling mills and steel works in Alabama: Nine contemplated and two projected. Of these, two have basic open-hearth steel plants, and two more are projected.

No steel was made in the state in 1894 or in 1895. The total amount made from 1888 to the close of 1893 will not exceed 4,000 gross tons.

Annual capacity of rolling mills, 173,300 gross tons with one mill not reporting. Allowing 10,000 gross tons for this one, the total annual capacity is 183,300 gross tons.

Production in 1895.

Number of completed rolling mills and steel works in the United States, January 1st, 1896, 505; annual capacity, double turn, 14,763,920 gross tons.

Forges and Bloomaries.

Anniston Bloomary, Cherokee Iron Company, Cedartown, Georgia. Works at Anniston, Calhoun county. Built in 1887; five forge fires and one hammer; steam power; product, blooms made from pig iron. Idle. Wm. C. Browning, President, and J. Hull Browning, Treasurer, 408 Broome St., N. Y.; J. R. Barber, Secretary and General Manager, Cedartown, Georgia.

Pipe Works, Car Wheel Works and Miscellaneous.

Bridge Building Works.

Southern Bridge Company, Birmingham. Works at Avondale, Jefferson county. Highway bridges. Annual capacity, 500 tons.

Gas and Water Pipe Works.

Anniston Pipe Works, Anniston Pipe and Foundry Company, Anniston, Calhoun county. Sizes from 3 to 30 inches. Daily melting capacity, 200 tons.

Chattanooga Foundry and Pipe Works, Chattanooga, Tenn. Works at Bridgeport, Jackson county. Sizes, from 14 to 36 inches, inclusive. Daily melting capacity, 125 tons.

Howard-Harrison Iron Company, Bessemer, Jefferson county. Sizes, from 3 to 60 inches, inclusive. Daily melting capacity, 300 tons.

Soil and Plumbers' Pipe Works.

Alabama Pipe Company, Bessemer, Jefferson county. Sizes, from 2 to 6 inches, inclusive. Daily melting capacity, 30 tons.

Birmingham Soil Pipe Works, Birmingham Soil Pipe Company, Birmingham, Jefferson county. Sizes, from 2 to 8 inches. Daily melting capacity, 10 tons.

Gadsden Foundry and Machine Works, Gadsden, Etowah county. Sizes, from 2 to 6 inches. Daily melting capacity, 10 tons.

Hercules Foundry, E. L. Tyler & Co., lessees, Anniston, Calhoun county. Sizes, from 2 to 12 inches. Daily melting capacity, 50 tons.

Car Axle Works.

Peacock's Iron Works, George Peacock, Selma, Dallas county. Iron and steel mine car axles. Annual capacity, 15,000.

United States (the) Car Company, Anniston, Calhoun county. Office, 1480 Old Colony Building, Chicago; 45 Broadway, N. Y. Car and locomotive axles. Daily capacity, 120.

Car Wheel Works.

Decatur Car Wheel and Manufacturing Company, New Decatur, Morgan county. Product, chilled, cast-iron wheels. Annual capacity 75,000. Removing to Birmingham.

Elliott (the) Car Company, Gadsden, Etowah county. Product, standard railroad car wheels. Annual capacity, 48,000.

Hood Machine Company, Birmingham, Jefferson county. Product, small tram wheels for mining cars. Annual capacity, about 12,000.

Peacock's Iron Works, George Peacock, Selma, Dallas county. Product, all kinds of small car wheels. Annual capacity, 35,000 self-oiling and 15,000 plate wheels.

Carbuilding Works.

Elliott (the) Car Company, Gadsden, Etowah county. Freight cars. Annual capacity, 3,600.

Peacock's Iron Works, George Peacock, Selma, Dallas county. Mine, logging and other small cars. Annual capacity, 5,000.

Union Iron Works Company, Selma, Dallas county. Logging, push, cane and other small cars. Annual capacity, 1,200.

United States (the) Car Company, Anniston. Offices, 1480 Old Colony Building, Chicago; 45 Broadway, N. Y.; works at Anniston and New Decatur. Annual capacity, 4,500 freight cars at each place.

CHAPTER VII.

PIG IRON; MARKET, GRADING, ETC.

THE PIG IRON MARKET; ITS EXTENT AND HOW TO IMPROVE IT.

BY

JAMES BOWRON, Birmingham, Ala.

(*Proc. Ala. Indust. and Sci. Soc. Vol. V, 1895, pp. 30-35.*)

The extent of the market for southern pig iron is very remarkable. According to the statistics published by Mr. Swank, of the American Iron and Steel Association, there were produced in the year 1894, 6,657,388 gross tons of pig iron, of which 120,180 tons were spiegeleisen and ferromanganese, leaving 6,537,208 tons to represent all of the iron produced in the United States for melting, or puddling. There were also produced in the same year 4,412,032 tons of steel of various kinds. The production of this steel would require more than a corresponding number of tons of iron, so as to allow for the waste incident to any method of conversion; but as this is a commercial and not a scientific paper, I am content to offset the above mentioned tonnage of spiegeleisen and ferro against the waste incurred in conversion, and simply to deduct the 4,412,032 tons of steel from the available tonnage of iron, which leaves 2,125,176 tons available for foundry and forge purposes. This may be considered to represent the American market during the restricted and unusual year of 1894, for the importations of foreign pig iron were only 15,582 tons, consisting mainly of Scotch, but including some Swedish pig iron.

PIG IRON ; MARKET, GRADING, ETC. 127

The production in Alabama in 1894 was 592,392 tons, and in Tennessee 212,773 tons, being 805,165 altogether. This production for Alabama represents 27.8 per cent. of the entire consumption of the United States, of pig iron apart from that used in the manufacture of steel; and for Tennessee a similar production of 10 per cent., being for the two States an aggregate of 37:8 per cent.

Without entering into any labored comparisons with other States or districts, this percentage is quite sufficient to show the commanding position taken by these districts in the general foundry and rolling mill trade, This is emphasized by the wide area over which the iron is distributed.

I give at this point the two following tables, showing the distribution by the Tennessee Coal, Iron & Railroad Company for the calendar year 1888, and for the twelve months ending July 1st, 1895:

1888.		1895.	
Tons of 2240 lbs.		Tons of 2240 lbs.	
Alabama	15,790	Alabama	$9,054
Arkansas	122	Arkansas	308
California	150	California	842
Colorado	1,200	Colorado	158
Connecticut	1,505	Connecticut	1,950
Delaware	36	Delaware	360
Georgia	18	Dist. of Columbia	293
Illinois	16,193	Florida	60
Indiana	9,599	Georgia	1,910
Iowa	2,300	Illinois	38,736
Kansas	383	Indiana	28,047
Kentucky	19,808	Iowa	2,251
Louisiana	588	Kansas	1,034
Maryland	268	Kentucky	48,376
Massachusetts	3,929	Louisiana	635

1888.

Michigan	16,582
Minnesota	70
Mississippi	155
Nebraska	234
New Jersey	1,820
New York	22,495
North Carolina	15
Ohio	65,561
Pennsylvania	6,777
Rhode Island	818
Tennessee	13,184
Texas	221
West Virginia	387
Wisconsin	818
Total Domestic	224,634

Foreign in 1895.

Canada	2,034
England	250
Italy	17
Nova Scotia	40
Mexico	357
	2,698

1895.

Maine	1,925
Maryland	4,748
Massachusetts	4,456
Michigan	28,170
Minnesota	612
Mississippi	560
Missouri	29,851
Nebraska	276
New Hampshire	295
New Jersey	31,965
New York	44,690
North Carolina	1,102
Ohio	136,487
Oregon	685
Pennsylvania	27,215
Rhode Island	645
South Carolina	57
Tennessee	30,368
Texas	1,605
Vermont	800
Virginia	580
Washington	50
West Virginia	18
Wisconsin	9,038
Total Domestic	570,212
Total Foreign	2,698
Total Domestic and Foreign	572,910

An examination of these tables shows the following interesting points:

1st. That the number of States consuming these brands of Southern iron increased in six years from 30 to 39, besides the addition of five foreign countries.

2d. That the home consumption had so far increased that Alabama moved up from 7th in progressive importance to 2nd, and Tennessee from 8th to 7th. This is a point of supreme importance, for notwithstanding the fact that Birmingham iron ranges from Mexico to Canada, and from San Francisco to Liverpool, it is obvious that distant markets can only be controlled by the sacrifice of profits, and that it is to the development of the home market, that can be reached without the payment of intervening freight charges, that we must look for our profitable business.

Obviously, therefore, everything that producers of pig iron in this district can do should be done to advance the interests of rolling mills, pipe works, machine foundries, &c., which are located beside us. We should advertise their products, give them our patronage, become personaly familiar with the character of their business, quality of the iron they use, their methods of treatment, the stocks they carry, and deliver to them the quantity and quality of iron that will subserve their necessities.

There are three ways in which the market for Alabama iron may be enlarged, namely: the development of (a) the home market; (b) the domestic; (c) the foreign.

(a) For the enlargement of the home market it is necessary for us to bring continually under the notice of manufacturers in other parts of the country the advantages which our cheap iron and coal, our mild climate, and reliable labor afford. This is doubtless done at present, with perfect loyalty to the district, by the gentlemen who are interested in it, but necessarily in a manner which is more or less desultory; and if we had a well-organized iron trade association the work might be done continuously and systematically. Many of the largest consumers of our iron have never been in the South, and their attention has not been personally directed to the consideration of the removal of their existing plants, or the establishment of new ones. A thoroughly intelligent representative of the district might be sent to call upon and make a presentation of our case to such consumers at a distance as might be selected by the association. Facts and figures in the same direction should be submitted in every case where a northern consumer of our iron is burned out and compelled to rebuild.

It is needless to say that the chief direction in which our efforts should be united, is to impress upon the producers of basic open hearth steel the advantage to accrue to them from the consumption of our metal at the point of production, where, if desired, it can be furnished hot. With the establishment of such a plant, works for the production of boiler plates, sheets for tinning, bars, structural and bridge work, wire rods, and railroad material will readily follow.

(b) For the enlargement of the domestic market, the most desirable thing to be done, in my judgment, is to secure uniformity in grading and naming iron, and selling it upon terms of uniformity. It is very unsatisfactory to the consumer in Canada or Minnesota to buy a car load of forge iron for foundry purposes, and next month to buy from another producer a car load of No. 2 soft and find that it contains less silicon and is less fluid.

It is scarcely too much to say that the whole question of grading iron is assuming a more complex condition, and if it is not in a somewhat chaotic state, the minds of some of the graders have attained that undesirable goal! Harassed by the pressure of evil times and the desire of consumers for something cheaper, the effort has been continually made not to split hairs, but to split grades in a corresponding degree of fineness. This leads to an absence of physical or chemical lines of demarcation, and makes the question of grading depend more than ever on the individual opinions of the maker and consumer, who naturally look at it from different standpoints, and arrive at different results. This leads to considerable friction, and in the long run Southern iron gets a bad name. With the organization, as before suggested, of a strong local trade association, the names of the grades could be definitely agreed upon, and arrangements made for at least monthly or bi-monthly interchange of visits from one works to another, so that the members might agree on the maintenance of a common standard, and correct discrepancies and divergencies from it.

(c) The question of developing a foreign market is one which at some future day will be of very great interest. When prices of Birmingham iron were at the lowest notch, on April 1st, 1895, it was then possible to put iron for export f. o. b. ships at tide water in Pensacola or Mobile Harbor, $2,00 per ton below the corresponding value of the cheapest English iron, and it was found practicable to lay down iron in Liverpool, grade for grade, at less than the price of Middlesboro iron shipped across England to that point. The facts also developed that at those figures we have for Alabama iron an exceedingly good chance for competition in all Mediterranean ports, very large quantities of iron being shipped from England to Barcelona; Genoa, Civita Vecchia, and other Italian ports.

This iron may be shipped in conjunction with steam coal or foundry coke. It would be an experiment of somewhat doubtful outcome as to whether the coke might, by the rolling of the vessel in an Atlantic voyage, be unreasonably broken up in comparison with the shorter voyage sustained by the English coke; but, if not, there is room for the shipment of Alabama coal and coke, in competition with English, into Mediterranean ports, although it is fair to say that there has not yet been a profit demonstrable in that business.

The main difficulty, however, in the development of a European market for our iron is and will be the commanding of marine tonnage; and any other business that could be grouped with pig iron, such as the exportation to Spain or Italy of coal or coke, or either, to Genoa, Bremen or Havre of cotton will materially facilitate the solution of the difficulty.

Whenever our prices for iron fall again so as to bring them below the parity of English figures, we should commence to work on ship brokers, and get in touch with both regular established lines of steamers, and also "tramp" steamers and sailing ships. The same remarks apply with as great force to the less importaht markets of Bombay, Calcutta, Melbourne, Yokohama and others. These markets are dominated, in pig iron, iron pipe, coal and coke, by the English because of the present exchange of commodities, which has settled steamers and sailing ships in certain marine lanes of travel, from which it will require patience and perseverance on our part to divert them.

GRADING COKE IRON.

The Grading of Birmingham Pig Iron.

BY

Kenneth Robertson, Birmingham, Ala.

(*Trans. Amer. Inst. Min. Engrs., 1888–1889, Vol. XVII, pp. 94–96.*)

All strangers visiting this district are struck with the

peculiar manner in which the pig iron is graded. There are eleven regular grades,' besides which, when gray forge is ordered, one-half of Nos. 1 and 2 Mill are shipped. Occasionally there is another grade known as Silver Mill, which is made so seldom that I can not describe it, and have no sample to exhibit.

Most of you have found it difficult to grade properly and uniformly under the simpler system which obtains elsewhere, and can consequently readily imagine the increased difficulty with us. Each furnace employs an "expert," and even with this precaution the system is not conducive, at all times, to amicable relations between buyers and sellers. I am told that it was adopted at the time Southern irons were seeking a market; but it still remains, although the time has come when our iron is sought for, and has obtained for itself a place in the markets of the country. The grades are as follows:

No. 1 Foundry, a large-grained, dark-colored iron with crystallization extending well out to the edges of the pig In my experience but little of it is made, and I am inclined to regard it as more of a freak than a product. An average of three analyses shows 3.66 per cent. of silicon in this grade.

No. 2 Foundry is the equivalent of a No. 1 Foundry at the North. An average of eighteen analyses gives 3.02 per cent. of silicon.

No. 2½ Foundry corresponds to No. 2 Foundry elsewhere. An average of eight analyses shows 3.02 per cent. of silicon.

No. 1 Mill is also known as No. 3 Foundry, and in it are included irons which are not good enough for 2½ Foundry, and also those which are equal to what is known as gray forge in the Lehigh Valley and vicinity. The best of this iron is used for foundry purposes. An average of four analyses gives 2.87 per cent. of silicon.

No. 2 Mill is between 1 Mill and Mottled, and contains 2.44 per cent. of silicon as an average.

No. 1 C is open-grained silver-gray. I have but one analysis, which shows 5.25 per cent. of silicon.

No. 2 C is close-grained silver-gray. Average of three analyses, 7.09 per cent. of silicon.

No. 1 Bright is a foundry iron which is light in color but open-grained. It is made by every furnace man in every district, at times; but it is only in this section that it is separated from the foundry irons. Elsewhere it would be shipped as No. 1 Foundry. Average of three analyses, 3.69 per cent. of silicon.

No. 2. Bright is one grade lower; is closer-grained; and the average of fourteen analyses is 3.11 per cent. of silicon. Elsewhere it would be a No. 2 Foundry.

To complete the number we have Mottled and White, which are the same here as elsewhere.

An idea has been prevalent for a long time that Southern irons are highly siliconized and weak; that the product of the furnaces is not foundry, but chiefly of lower grades; and that the lower grades are sold with difficulty.

The preceding analyses show that the foundry-irons do not contain more silicon than irons of the same grade in other districts; the mill irons are higher in silicon than those of Glendon and Andover, but are sold without difficulty. As to the product of the furnaces, I will give the percentages of each grade which one of the furnaces under my charge made during ten months' working under very disadvantageous circumstances.

Other furnace in the district have undoubtedly done much better; and these figures are not given as typical, but merely to show that we do make foundry iron, and that the greater portion of our product is not of lower grades. The average of twenty-seven determinations of phosphorus is 0.66 per cent.

Percentage of Grades made:

PIG IRON; MARKET, GRADING, ETC. - 135

No. 1	Foundry	0.27
No. 2	"	26.23
" 2½	"	19.48
" 1	Mill	33.82
" 2	"	6.85
" 1	C.	0.56
" 2	"	1.39
" 1	Bright	6.07
" 2	"	2.76
Mottled		2.38
White		0.19
		100.00

Calling the Bright irons foundry, which they are, the proportion of foundry-iron made was 54.81 per cent.; probably half the 1 Mill would have been classed as foundry elsewhere. The results are not considered as the *ne plus ultra* of furnace work, but will show what we are doing, and also that the impression that but little foundry iron is made here is erroneous,

(This paper was prepared for the Birmingham meeting of the American Institute of Mining Engineers, May, 1888. At that time the prices, f. o. b. furnace, Birmingham district, for the various grades given were about as follows, per ton of 2,240 lbs.:

No. 1	F	$14.50
" 2	F	14.00
" 2½	F	13.90
" 3	F	13.75
" 1	M	12.50
" 2	M	10.50
" 1	C	
" 2	C	
" 1	Bright	15.00

" 2 Bright..................... 11.75
Mottled........................... 11.00
White 9.00

The freight rates to various important points were as follows, car load lots: Cleveland, $4.00; Cincinnati, $2.75; Chicago, $4.00; Columbus, O., $3.15; Boston, $3.86; New York, $3.86; Phila., $3.86; San Diego, Cal., $21.87; Wheeling, $4.50; Moline, Ill., $5.12; Greencastle, Ind., $3.40.

W. B. P.)

THE PROPER GRADING OF SOUTHERN PIG IRON

BY

A. E. BARTON, ENSLEY, ALA.

(*Proc. Ala. Indust. & Sci. Soc., Vol. III, 1893, pp. 35-39.*)

When requested by our president to prepare a paper to read at the present meeting of the Society, I felt that I had not sufficient time at my disposal for such a task; but was told that all that was necessary was to bring forward some subject for discussion..

The question as to the proper grading of Southern Pig Iron is an important one in these times of depression, and it is necessary that the maker should take into consideration the special needs of the customer more than has hitherto been done.

Many consumers are now calling the chemist to their aid, and look more to the chemical composition of the pig than to fracture. Some, however, still remain in the old groove, and are guided altogether by the strength of the iron and appearance of the fracture when broken, which guide is often misleading.

If we look back some twenty-five years, we find many firms in Scotland and in the Middlesboro district of England, offering only two grades of iron—Foundry and Forge—the iron all being shipped "long," and only an occasional pig being broken. The requirements of customers soon called for an extension of these grades, and six grades were recognized—three of Foundry, i. e., 1, 2, and 3; and three of Forge, i. e., Gray Forge, Mottled and White. Under certain conditions in the working of the furnaces, a light colored weak iron was made, which the furnace men called "bright iron." This iron was at first very little in demand, and when sold brought a very much lower price than the Foundry grades proper.

Occasionally, when the stock used was of inferior quality and the fuel consumption high, a still weaker iron would be made, which, from its peculiar fracture, was called "glazed pig," and was generally put back into the furnace as being unsaleable. It was noticed that when this pig was used as scrap in the furnace, the quality of the iron made seemed to improve and become more open in grain. Prof. Turner, of Birmingham, England, gave this matter considerable attention, and discovered that when a certain proportion of the glazed pig was mixed with a hard iron and melted in a cupola, it had a tendency to open the grain and make the iron softer, and that "bright" iron brought about the same result in a modified degree. After another research, it was found that the large amount of silicon contained in the "bright," and "glazed" pig was responsible for the result, and it found that by the use of these irons a considerable amount of scrap could be used in foundry mixture which, by repeated melting, had become too hard to use alone, and only in small proportions when mixed with foundry iron. After this, "bright," and "glazed" pig found a ready sale, and several grades of bright and silvery iron were established.

Some six or seven years ago there were fifteen recognized grades of Southern iron, as follows : Open silvery; close silvery ; open bright ; medium bright ; close bright; 1 foundry; 2 foundry; 2½ foundry; 3 foundry ; extra 1 mill ; 1 mill ; 2 mill ; silvery mill ; mottled ; and white.

About five years ago it was decided, at a meeting of Southern Iron Masters, that a revision of grades was necessary, and that the South had too many grades, and the following change was made ; Open, and close silvery were continued and called silver gray; open, medium and closed bright were condensed into two grades and called Nos. 1 and 2 soft; No. 2 foundry was called No. 1 foundry, and the old No. 1 foundry, which was a very open iron and seldom made, was mixed in with the old No. 2 foundry ; No. 2½ and No. 3 foundry were mixed together and called No. 2 foundry ; extra 1 mill became No. 3 foundry ; Nos. 1 and 2 mill were continued and called gray forge; silvery mill was no longer recognized as a grade. Mottled and white remained the same.

This alteration in the classification caused a good deal of confusion and many complaints from customers. Old buyers of Southern iron complained that the silver gray shipped was "mixed," and to any one grading by fracture alone it certainly looked mixed. The two irons, however, are practically identical in composition chemically, the close flaky iron generally running slightly the highest in silicon, which will vary from 4 to 5½ per cent., and both are somewhat low in total carbon.

To meet the wishes of a certain class of customers, the old method of grading silver gray has gradually been adopted by producers, and we have now two grades of silvery iron recognized, Nos. 1 and 2, corresponding to the old open, and close. In soft irons, the openest pigs of medium bright were thrown into 1 soft, and the remainder called 2 soft. The latter can not be graded so uniformly as to fracture as could be desired, for this reason, and is considered by many buyers as an off grade.

Soft iron should contain from 3 to 4 per cent. of silicon, and be practically free from sulphur, whilst the carbon, though not so high as in the foundry grades, runs higher than in silver gray, combined carbon being usually about the half of 1 per cent., and graphite 2 to $2\frac{1}{4}$ per cent. in No. 1 soft, and $\frac{3}{4}$ of one per cent., and $1\frac{1}{2}$ per cent., respectively, in No. 2 soft.

Soft iron is used as a softener in mixtures, and to use up scrap, and is essentially siliceous. One mistake, often made by the grader, is to class as No. 2 soft the pigs from a foundry cast that have been chilled during their course down a long runner, and have, from this cause, a light colored appearance, with a close edge. These pigs generally run about 2 per cent. in silicon, and should be graded either as 2 or 3 foundry. Care should also be taken with the grading of 1 soft, for the same reason. Recently a pig was graded by those competent graders as 1 soft, by fracture alone, without seeing the pig broken, and on analyses it showed $1.12\frac{1}{2}$ per cent., silicon, and was not a soft iron at all. Many buyers, however, would have rejected a car of such iron on sight, if shipped to them as 2 foundry, and would have used it as 1 soft, probably with disastrous results.

The grading of the three straight foundry grades does not require much comment. The standard amount of silicon in each grade should be about as follows: 1 foundry, 2.75 per cent.; 2 foundry, 2.50 per cent.; and 3 foundry, 2.00 per cent.; and these contents should be maintained as nearly as possible by repeated analyses and changes in the burden of the furnace, when necessary. It was in forge iron that the change in the grading caused the greatest trouble. Until lately, the furnaces of the district made sufficient gray forge iron, in endeavoring to make foundry irons, to meet all demands, and the forge iron thus made was apt to be high in silicon, and very wasteful for rolling mill iron, though suitable as a mixture in pipe works and foundries, and complaints from rolling mills that had been using No. 2 mill came in thick and fast. Pipe works would also get a car of said iron occasionally, and the furnace would generally hear about it.

Graders soon saw the impracticability of having only one grade of Gray Forge, and tacitly made two grades in their yard, though only one was recognized; ascertaining before making shipment if the Gray Forge was to go to rolling mill or foundry, and shipping accordingly 2 mill, or 1 mill. These two grades are now generally recognized, being called Gray Forge, and Foundry Forge, the former being a much harder iron, lower in silicon and higher in combined carbon, with a different crystal.

As blast furnace practice inproves in the South, and the iron making materials are more carefully selected, the furnaces will be kept more steadily in one grade of iron, and the iron will become, and is now becoming, more improved; and should the demand justify it, the furnaces will be run with the express purpose of making low silicon Gray Forge, similar to the Northern practice.

A considerable portion of this paper will be ancient history to many here, but it brings forward a very important subject—a subject which until lately has hardly been given the attention it deserved—that of giving customers uniform iron, and iron best suited to their special needs, and removing the stigma that a buyer of Southern iron never quite knows what he is going to get.

(This paper was prepared for the Birmingham meeting of the Alabama Industrial and Scientific Society, May, 1893. At that time the prices, f. o. b. furnace, Birmingham district, for the grades given, were about as follows, per ton of 2240 lbs.:

Silver Gray $9.50
1 Soft 9.00
2 Soft 8.75
1 Foundry 10.75
2 Foundry 9.50
3 Foundry 8.50
Gray Forge 8.25
Mottled 7.90
White 7.50

The freight rates to inportant points, car load lots, were as follows: Atlanta, $1.30; Boston, $4.36; Buffalo, $4.40; Cleveland, $3.85; Cincinnati, $2.75; Detroit, $3.49; Erie, $4.40; Evansville, $2.75; Louisville, $2.10; New Orleans, $2.55; New York, $4.31, Philadelphia, $4.31; St. Louis, $3.25; Worcester, $5.14.

The agreement between the Southern coke iron makers, to which Mr. Barton alludes, was made during the summer of 1888, and was as follows, according to the circular that was issued:

"CHANGE OF NOMENCLATURE OF SOUTHERN COKE IRON GRADING.

The system of grading pursued by Southern coke furnaces in the past having been peculiar to the district and out of line with the grading followed by Northern furnaces, it has been determined to change the nomenclature of Southern grading as to make it conform to to the standard throughout the country.

Therefore, on and after October 1st, 1888, the following schedule will be uniformly pursued by the companies whose signatures are attached:

No. 1 Foundry, same as hitherto called No. 2.
No. 2 Foundry,   "   "   "   No. 2½.
No. 3 Foundry,   "   "   "   No. 1 Mill.
No. 1 Soft,   "   "   "   Open Bright.
No. 2 Soft,   "   . "   "   Close Bright.
Silver Gray,   "   "   "   Silver Gray.
Gray Forge,   "   "   "   No. 2 Mill.
Mottled,   "   "   "   Mottled.
White,   "   "   "   White.

Sales agents are required to invoice in accordance with this schedule all shipments on new orders, and also those on orders taken before this goes into effect and not yet completed.

Tenn. Coal, Iron & Ry. Co., operating the 4 Ensley, 2 Alice, 3 South Pittsburg and 1 Sewanee furnaces.

Sloss Iron & Steel Co., operating the 4 Sloss furnaces.

Nashville Iron, Steel & Charcoal Co., operating the 2 Nashville furnaces.

Williamson Iron Co., operating the 1 Williamson furnace.

Mary Pratt Furnace Co., operating the 1 Mary Pratt furnace.

Roane Iron Co., operating the 2 Rockwood furnaces.

Citico Furnace Co., operating the 1 Citico furnace.

Dayton Coal & Iron Co., L'd., operating the 2 Dayton furnaces.

Gadsden-Alabama Furnace Co., operating the 1 Etowah furnace.

Walker Iron Co., operating the 1 Rising Fawn furnace.

Chattanooga Iron Co., operating the 1 Chattanooga furnace.

Sheffield & Birmingham Coal, Iron & Ry. Co., operating the 3 Cole furnaces.

Eureka Co., operating the 2 Eureka furnaces.

Woodward Iron Co., operating the 2 Woodward furnaces.

DeBardeleben Coal & Iron Co., operating the 2 DeBardeleben furnaces."

This was the first concerted step taken towards uniformity of grading in Southern Coke Iron, and was productive of very considerable benefit to the trade.

W. B. P.)

THE GRADING OF SOUTHERN COKE IRON WITH SPECIAL REFERENCE TO THE BIRMINGHAM DISTRICT.

(*Proc. Ala. Indust. & Sci. Soc., Vol. VI, 1896, pp. 11-14.*)

BY

W. H. BRANNON, Bessemer, Ala.

Eight years ago there were in the Birmingham District 15 grades of iron, viz :—1 Foundry ; 2 Foundry ; 2½ Foundry ; 3 Foundry ; Extra No. 1 Mill ; No. 2 Mill ; Mottled ; White ; No. 1 Bright ; Medium Bright ; Close Bright ; No. 1 Silvery ; No. 2 Silvery ; and Silvery Mill.

This list was revised in 1888, and to-day we recognize 11 grades, viz :—No. 2 Silvery ; No. 1 Silvery ; No. 2 Soft ; No. 1 Soft ; No. 1 Foundry ; No. 2 Foundry; No. 3 Foundry ; No. 4 Foundry ; Gray Forge ; Mottled ; and White.

In 1888 very little attention was paid to chemical analysis, the irons being graded almost entirely by color and granulation. In addition to having a fair knowledge of the principal chemical ingredients of pig iron the grader now must be thoroughly familiar with the four points in uniform grading, viz :—color, granulation; fracture and face.

No. 2 Silvery contains from 5 to 5.50 per cent. of silicon, has very little or no granulation, and is almost smooth, with a galvanized appearance. No. 1 Silvery has some granulation, and a smooth face, and contains from 4.50 to 5 per cent. of silicon. Both these irons are weak in fracture, and show a fine, silvery lustre on a fresh face, and are flaky. They should exhibit no dark spots, and the crystallization is obscure. They are what they purport to be 'Silvery irons,' and the difference between them, on the yard, is mainly, one of granulation. They are the hottest irons, and contain much more silicon and much less combined carbon than any of the other grades. Their carbon is almost wholly in the shape of graphite, but the large excess of silicon prevents this ingredient from conferring a dark color on the iron.

No. 2 Soft contains 3.50 to 4.0 per cent. of silicon. No. 1 Soft from 3.0 to 3.5 per cent. They are both of a light color, smooth face and weak fracture. A distinct granulation begins to be apparent in No. 1 Soft, which is more pronounced in No. I Soft, but in neither of these grades is the granulation so marked as in the Foundry irons.

The Soft irons are darker than the Silvery irons, but lighter in color than the Foundry irons, and the granulation is not so jagged as in these latter grades. In particular they do not show a silvery appearance, and are not flaky. The increasing ratio of graphite to silicon begins to manifest itself in the Soft irons in the darkening of the color as compared with the Silvery irons.

No. 1 Foundry contains from 2.50 to 3.0 per cent. of silicon, has a very open and regular granulation extending through the entire face, and a dark gray color. The crystallization is marked, and the face is rough to the feel. The difference between this and No. 2 Foundry, which contains from 2.25 to 2.50 per cent. of silicon, is the same in kind as exists between the two silvery, and the two soft irons, and is chiefly one of granulation. In No. 2 Foundry the grain is not so open as in No. 1 Foundry, nor is the crystallization so coarse. The color may be as dark in one as in the other, but in No. 1 Foundry there is a deep blackish gray color which is absent in No. 2 Foundry.

No. 3 Foundry contains from 2.0 to 2.25 per cent. of silicon, and resembles No. 1 and No. 2 Foundry in structure, but the granulation is much less marked. The crystallization is finer than in No. 2 Foundry, and the color, while still dark gray, is not so pronounced.

No. 1 Foundry, recently called Foundry Forge, shows the dark gray color of the other foundry irons, but the granulation is closer and the crystallization finer. It carries from 1.75 to 2.0 per cent. of silicon. Taken together the foundry irons are distinguished by dark gray color, open grain, and well marked crystallization, three points which are seen to the best advantage in No. 1 Foundry.

Gray Forge is the old No. 2 Mill. It has 1.50 to 1.75 per cent. of silicon, and shows a pebbled granulation in the center, with mottled edges about one-quarter of an inch deep all around. It has a blistered and pitted face, and is frequently honey-combed on the fractured end, some of the holes being an eighth to a half an inch deep.

Mottled iron has from 1.25 to 1.50 per cent. of silicon, shows no granulation, and has a pepper-and salt appearance on a fresh face. It begins to show an increasing amount of combined carbon, about one-half of the total carbon being in this condition.

White iron has from 1.0 to 1.25 per cent. of silicon, shows no granulation, and is often as white as bleached linen. It carries very little graphite, and is usually high in sulphur. It is very hard, often resisting the drill, and on this account is difficult to sample properly.

In sampling pig iron one of two methods may be used, the choice depending on the extent of the subsequent analysis. When silicon, sulphur, phosphorus, manganese and total carbon are to be determined the iron is best sampled from the runner, from 4 to 6 small ladles-full being taken during the cast and shotted in a bucket of cold water. When graphite, and combined carbon are also to be determined boring must be resorted to. In this case two methods may be used. In the first, the face of the pig is bored in three places to the depth of $\frac{1}{2}$ to 1 inch along a line drawn diagonally across the face, the borings being mixed. In the second, the pig is bored diagonally almost entirely through in one place.

In boring pig iron care must be taken to prevent the intermixture of sand from the pig with the borings, and it is well to put a careful man in charge of the drill. In boring chilled pig, and in sampling from the runner, there is, of course, much less danger of adhering sand getting into the borings. A neglect of this matter may often mislead the grader, as sand in the borings shows up as silicon in the pig, and a No. 3 Foundry may be classed as a No. 1 Soft. It is a difficult and tiresome matter to separate sand from borings by means of a magnet, and at the best entails a good deal of extra and unnecessary labor upon the chemist.

The tendency of the trade is now strongly towards a closer chemical inspection of the irons offered for sale, and the grader who intends to keep up with his profession must take this fact into consideration. He must, therefore, acquaint himself with the effect of the chief constituents upon the various irons in respect of color, granulation, fracture and face. He is called upon every day to decide questions involving a great deal of money, and as it sometimes happens that he can not wait for an analysis he must be prepared to grade without it. But he should by all means cultivate the closest intimacy with the laboratory, and have the grades analysed as often as possible, and not neglect to inform himself as to the influence of the burden, heat and pressure upon the product under his care.

(This paper was prepared for the Birmingham meeting of the Alabama Industrial and Scientific Society, May, 1896.

At that time the prices, f. o. b. furnace Birmingham district, for the grades given were about as follows, per ton of 2,240 lbs.:

 Open Silver Gray..................$8.75
 Close " " 8.50
 1 Soft............................ 7.75
 2 Soft............................ 7.50
 1 Foundry........................ 8.25
 2 Foundry 7.75
 3 Foundry 7.25
 4 Foundry 6.90
 Gray Forge....................... 6.75
 Mottled 6.75
 White............................ 6.25

The freight rates to various important points were as follows:

FREIGHT TARIFF FOR PIG IRON,

In Effect February 24th, 1896.

Birmingham To—	Distance in Miles.	Rate per Ton. Car load not less than 17½ tons of 2,288 lbs.	Route.
Atlanta............	167	$ 1.30	Rail.
Baltimore..........	1050	3.60	Rail and water.
Boston	1450	4.10	"
Buffalo	950	4.40	Rail.
Cincinnati.........	504	2.75	"
Cleveland	767	3.90	"
Chicago............	650	3.85	"
Detroit	766	3.95	"
Galveston	800	4.83½	"
Hamilton, Canada..	975	5.10	"
Louisville..........	394	2.50	"
Mobile	276	2.50	"
New Orleans.......	417	2.50	"
New York.........	1225	3.75	Rail and water.
Norfolk............	765	3.00	"
Philadelphia.......	1150	4.75	"
Pittsburg..........	817	4.40	Rail.
Portland, Oregon..	3675	14.47	"
San Francisco......	2900	13.34	"
Savannah..........	448	2.90	"
St. Louis...........	528	3.25	"
Toronto, Canada...	996	5.10	"

W. B. P.)

OBSERVATIONS ON GRADING COKE IRON.

(Proc. Ala. Indust. & Sci. Soc., Vol. VI, 1896, pp. 15–23.)

BY

William B. Phillips, Birmingham, Ala.

The grading of coke iron calls for a considerable amount of skill, which can be attained only by experience and the closest attention to details. Even then it not infrequently happens that the best graders will be at fault and the question can be settled only by the chemist.

Reference has been made by Mr. Barton (Proc., Vol. III, 1893, pages 35-39) to a case where an iron which was afterwards found to contain 1.12 per cent. silicon was graded by these different men as No. 1 Soft. Other instances might be adduced in support of the proposition that, after all, the proper place for grading iron is the laboratory. It is not meant by this that iron should be graded only in the laboratory, for this would entail considerable expense, which might not be warranted by the situation. But that the closest affinity should exist between the grader and the chemist, no one who has observed the course of the Southern coke irons during the past eight or nine years can reasonably deny.

When Mr. Robertson prepared his paper on the grading of Southern coke irons (Trans. Amer. Inst. Min. Engrs., Vol. XVII, 1888-89, pp. 94-96), the foundry irons carried more than 3 per cent. of silicon, and there was constant complaint by consumers that there was too much variation in the grades. This was to a great extent allayed by the practice that began with the circular issued by the Southern iron men in 1888, which has been quoted in full in my Report on Iron Making in Alabama.

But it was found in 1893 that another grade was needed, intermediate between 3 Foundry and Gray Forge, and it was established by some companies under the term Foundry Forge. To this the objection was made that the term was self-contradictory, an iron could not be at once Foundry, and Forge. In 1895 the grade was abolished, and what was formerly Forge is now termed No. 4 Foundry.

Now and then it happened that a close, fine grained silvery looking iron would show on analysis not more than 2 per cent. of silicon, while again, without greatly altering in appearance, it would show from 2.90 to 3.10 per cent. silicon. If the silicon was about 2 per cent. the iron was termed Foundry Forge, as it is now termed

No. 4 Foundry; if the silicon was about 3 per cent. it was and is yet, termed No. 1 Soft.

Ordinarily and when grading for the same furnace running on about the same burden, the competent grader comes very near the proper grade, and can be trusted to ship on his own judgment. But when complaints arise, as they do sometimes, and especially on a depressed market, the only way in which the ire of the consumer can be placated is to show him that the iron he objected to as not being No. 3 Foundry, for instance, does really contain from 1.75 to 2 per cent. of silicon, and falls within the limits for this particular grade. This much as to silicon. But how is it in respect to graphitic, and combined carbon? Is the iron to be graded solely by its silicon content? It is granted that for the most part iron can be fairly well graded on its content of silicon, and that the variation of this element confers upon the iron peculiarities of color, granulation, fracture, and face that are more strongly marked than peculiarities due to other elements. It is this fact that has rendered possible the present system of visual and tactual grading. It was quietly assumed that if the silicon was all right, the iron was all right, and this was supplemented by the further assumption that if the iron was all right the silicon was all right. In this way it was possible even for a conscientious grader to fortify himself behind a pretty high wall of silicon, and fire silicon bombs *ad libitum*.

The easiest way of grading iron is by its silicon content, but it by no means follows that it is the best way, or the only way. Leaving out the content of sulphur, as not seriously affecting any of the grades above Gray Forge, there should be certain ratios established between silicon and combined carbon for the Soft and Foundry irons. The variation in the amount of silicon does, of course, influence the quality of the iron, and one might go even farther and allow that it influences the iron more than any other single element. But combined carbon is by no means to be neglected.

In 29 complete analyses of iron graded as No. 3 Foundry, I found that the silicon varied from 1.45 to 3.83 per cent., the average being 2.37 per cent. Five of the samples should have been graded as No. 1 Soft, as the silicon was between 3.04 and 3.17 per cent., and one should have been No. 2 Soft with silicon 3.83 per cent. These irons were all graded on the yard by a careful and competent man, yet in 6 cases out of 29, or 20.7 per cent., the iron graded as No. 3 Foundry was really Soft. Excluding these six, the average silicon in the other 23 was 2.16 per cent., a result not far wrong, if at all, as No. 3 Foundry may vary from 1.90 to 2.20 per cent. of silicon. In the six cases in which the silicon was over 3 per cent. the combined carbon was 1.04 per cent., and the 23 others it was 0.82 per cent., the average of the 29 being 0.87 per cent.

The combined carbon in No. 3 Foundry does not usually run as high as 0.82 per cent., the average being about 0.40 per cent. In the Soft irons it should not be above 0.40 per cent., but in some cases especially when the iron resembles No. 3 Foundry, it may go to 1.00 per cent.

We have then to discriminate between Soft irons with over 3 per cent. of silicon, and the normal amount of combined carbon, and irons which contain over 3 per cent. of silicon and upwards of 1 per cent. of combined carbon. Grading on fracture and appearance some of these latter irons would be put in No. 3 Foundry; grading on silicon content they would go in the Soft irons, with the understanding that the combined carbon was abnormally high.

The same principle holds good in respect of the other Foundry irons, although in a less degree. It is this tendency of the lower grades of Foundry iron to show higher percentage of combined carbon than is usually the case that renders grading by fracture and appearance somewhat uncertain. In case of doubt a silicon estimation will enable one to decide whether or no the iron should be put in the Soft grades, and an estimation of combined carbon will show whether or no it should be stated that this element is above the average.

In a paper read before the Alabama Industrial and Scientific Society in 1895, which we have quoted in full, Mr. James Bowron said: "For the enlargement of the domestic market, the most desirable thing to be done, in my judgment, is to secure uniformity in grading and naming iron, and selling it upon terms of uniformity. * * * It is scarcely too much to say that the whole question of grading iron is assuming a more complex condition, and that if it is not in a somewhat chaotic state, the minds of some of the graders have attained that undesirable goal. Harassed by the pressure of evil times and the desire of the consumer for something cheaper, the effort has been continually made not to split hairs, but to split grades in a corresponding degree of fineness. This leads to absence of physical or chemical lines of demarcation, and makes the question of grading depend more than ever on the individual opinions of the maker and consumer, who naturally look at it from different standpoints, and arrive at different results. This leads to considerable friction, and, in the long run, Southern iron gets a bad name."

Mr. Bowron's long and intimate acquaintance with the commercial aspects of grading qualifies him to speak *ex cathedra*, and if he can deliberately take the position that uniformity in grading and naming iron is the most desirable thing that can be done towards enlarging the domestic market for pig iron, surely it is time to discuss the matter from every point of view, with the hope of arriving at some more reasonable system than is at present used.

The multiplication of grades may go on indefinitely according as the fancied needs of consumers increase in number. If a manufacturer asks for an iron carrying not more than 3.50 per cent. and not less than 3 per cent. of silicon with combined carbon not over 0.50 per cent., he should be able to get it.

There has recently been completed an agreement between the chief producers of Alabama coke iron whereby certain uniform prices for standard grades are to be observed. It is a very good thing as far as it goes, but it does not go far enough, nor strike very heartily at the root of the trouble.

The main point is to secure uniform grading, and this can certainly not be gained merely by establishing uniform prices. Mr. Bowron was unquestionably right in saying that uniformity in price *and uniformity in grading*, (the italics are ours) must be maintained if the domestic market is to be enlarged.

The local trade association of which he speaks could take the matter in hand, but a simpler and it seems to us a more satisfactory plan would be for the companies that made the agreement as to prices to make a similar agreement as to grading, and put a competent man in charge of it. The price depends upon the grading. It is not enough for the iron-masters to meet and say what the names of the grades shall be, nor to fix the price at which the grades thus named shall be sold. Unless there is at the same time an agreement as to what kind of iron shall be deemed No. 1 Soft, or No. 3 Foundry, the proctocol as to uniform prices is to a large extent abrogated. It is sure to happen that permission to ask a special price for a special iron will be solicited, and unless it is known what this iron is, what relation it bears to the grades whose prices are already fixed and agreed upon, how can there be any thing but confusion? One may say: "I am making an iron, or I have it and it is now piled, which to all ordinary grading would be put in No. 2 Foundry. But it carries less than 1.50 per cent. of silicon and is therefore not a typical No. 2 Foundry and I wish to ask a special price for it." He has called in his chemist and knows that the iron is not No. 2 Foundry, although it closely resembles it in granulation, color, fracture and face. He wishes to sell it on analysis, for this is really the gist of the whole matter.

By all means let there be uniform prices, but if the grading is not uniform what do the uniform prices amount to, after all? They are simply grade-splitters, and will inevitably lead to more confusion than at present exists, if they are not based on the chemical analysis of the irons.

Some people are inclined to regard the chemical grading of pig iron as a sort of Panjandrum, or Mysterious Monster, lying in wait for the unwary, and they begin to tell their beads as soon as a chemist heaves in sight. But no chemist who understands the situation in Alabama can declare out and out for laboratory grading, as no chemist can doubt that the present system is out of date, illogical, and cumbersome.

The purposes to which pig iron is put depend absolutely upon its composition; the color, fracture, granulation, and face have nothing to do with it except in so far as they indicate the existence of certain ingredients whose actual percentages can be determined only by the chemist. As regards grading the inferences to be drawn from data obtained on the iron yard are reliable only if confirmed by laboratory tests, and it is particularly ungrateful in graders and furnace managers to decry the further application of the very science upon which they base the practice of their art.

What changes are to be suggested? First the maintenance of a chief grader, whose business it should be to regulate the grading under conditions imposed by the separate companies. Second, the establishment of a central laboratory devoted to pig iron analyses. Third, the diminution of the number of grades and the substitution therefor of not more than six grades, differentiated by the content in silicon, and combined carbon, and possibly sulphur. These six grades might be as follows;

	Silicon.	Combined Carbon.	Sulphur.
Silvery Irons,	5 to 6	0.10 to 0.30	0.01 to 0.04
Soft Irons,	3 to 5	0.20 to 0.60	0.01 to 0.05
Foundry Irons,	2 to 3	0.30 to 0.90	0.01 to 0.07
Gray Forge,	1 to 2	0.40 to 1.25	0.04 to 0.09
Mottled,	0.6 to 1	0.50 to 1.80	0.06 to 0.11
White,	0.1 to 0.6	1.00 to 2.50	0.08 to 0.30

This scheme, or some modification of it in line with its general provisions would retain the present nomenclature, and bring it into closer accord with laboratory results. It would do away with five grades, which are no more than side-grades at best, and would enable the grader to exercise better discretion in the yard. The rapidity and accuracy with which the estimation of silicon, and combined carbon can now be made render it possible to have the results from the cast-house by the time the iron is ready to break and pile. The estimation of silicon now leaves very little to be desired, and while the estimation of combined carbon in pig iron is not so accurate as in steel it is sufficiently so for the purpose in hand. If objection be made to such a radical change much could be done to improve the present system without decreasing the number of grades, or interfering with the nomenclature. If a systematic record of the pigs sampled were kept it would be possible to control the grading within narrower limits than now maintain. An excellent system has been devised by consultation with Mr. W. H. Brannon, chief grader for the Tennessee Coal, Iron & Railroad Co., Mr. W. J. Sleep, manager of the American Pig Iron Storage Warrant Co., in this district, and two well known pig iron brokers whose names it is not necessary to mention. It was our purpose to have a convenient envelope prepared for holding the borings, and on the front of it we had the following:

Tennessee Coal, Iron & Railroad Co.

——Tons.   Grade——
No——Furnace.   Division——
Made——189—   Sampled——189—

Color $\begin{cases} \text{Light.} \\ \text{Dark.} \end{cases}$

Fracture $\begin{cases} \text{Weak.} \\ \text{Strong.} \end{cases}$

Granulation. $\begin{cases} \text{Regular.} \\ \\ \text{Irregular.} \end{cases}$ $\begin{cases} \text{Fine.} \\ \text{Medium.} \\ \text{Coarse.} \end{cases}$

Face. $\begin{cases} \text{Smooth.} \\ \text{Pitted.} \\ \text{Blistered.} \end{cases}$

Chilled edge——
Signed——

On the back of the envelope was printed the following:

Charges——

Burden.		Pounds.
Hard Ore............................	———	
Soft Ore............................	———	
Brown Ore...........................	———	
Stone. $\begin{cases} \text{Limestone.} \\ \text{Dolomite.} \end{cases}$	——— ———	
Coke...............................	———	
Total............	———	

To be taken before each cast.

	Time.			Average.
Revs. of Engine.				
Heat.				
Pressure.				

The line on the front of the envelope that does not apply to the sample is marked out, thus if the color is dark the word 'light' is marked out, if the granulation is regular and fine, the words irregular, medium, and coarse are marked out, if there is a chilled edge $\frac{1}{8}$ or $\frac{1}{4}$ inch deep it is so stated, if there is no chilled edge, the words are erased.

By the use of this envelope it is possible to have recorded a complete history of the sample under examination. The results reached are of the highest importance if the system is faithfully adhered to, for at any time, by reference to the laboratory books, it can be known what was the exact composition of any grade of iron, its physical peculiarities and the burden on which it was made.

We are convinced that if this, or a similar, system were intelligently and persistently followed the complaints of lack of uniformity in grading our coke irons would gradually disappear.

It is acknowledged on every side that the irons are *not* uniformly graded, and unless some steps are taken to remedy this most serious obstacle to the enlargement of our markets we shall always be met with the assertion that we are not doing what we could do to correct the evil.

TABLE XI.

Production of Iron Ore, Coal, Coke and Pig Iron in Alabama.

Year	Iron Ore. Tons of 2,240 lbs.	Coal. Tons of 2,000 lbs.	Coke. Tons of 2,000 lbs.	Pig Iron. Tons of 2,240 lbs.		
				Coke.	Charcoal.	Total.
1870	11,350	10,999				
1871		20,000				
1872	22,000	30,000			11,171	11,171
1873	39,000	44,800			19,895	19,895
1874	58,000	50,400			29,342	29,342
1875	44,000	67,200			22,418	22,418
1876	44,000	112,000		1,262	20,818	22,080
1877	70,000	196,000		14,643	22,180	36,823
1878	75,000	224,000		15,615	21,422	37,037
1879	90,000	280,000		15,937	28,563	44,500
1880	171,139	380,000	60,781	35,232	33,693	68,925
1881	220,000	420,000	109,033	48,107	39,483	87,590
1882	250,000	896,000	152,940	51,093	49,590	100,683
1883	385,000	1,568,000	217,531	102,750	51,237	153,987
1884	420,000	2,240,000	244,009	116,264	53,078	169,342
1885	505,000	2,492,000	301,180	133,808	69,261	203,069
1886	650,000	1,800,000	375,054	180,133	73,312	253,445
1887	675,000	1,950,000	325,020	176,374	85,020	261,394
1888	1,000,000	2,900,000	508,511	317,289	84,041	401,330
1889	1,570,000	3,572,983	1,030,510	608,034	98,595	706,629
1890	1,897,815	4,090,409	1,072,942	718,383	98,528	816,911
1891	1,986,830	4,759,781	1,282,496	717,687	77,985	795,672
1892	2,312,071	5,529,312	1,501,571	835,840	79,456	915,296
1893	1,742,410	5,136,935	1,168,085	659,725	67,163	726,888
1894	1,493,086	4,397,178	923,817	556,314	36,078	592,392
1895	2,199,390	5,705,713	1,444,339	835,851	18,816	854,667

TABLE XII.

FREIGHT TARIFF FOR COAL AND COKE. IN EFFECT FEBRUARY, 1896.

BIRMINGHAM To—	Distance in Miles.	Rate per Ton. Car load not less than 23 tons of 2,000 lbs. Local.
Atlanta	167	$ 1 05
Augusta	338	2 05
Charleston	476	2 05
Columbia, S. C	423	2 20
Columbus, Miss	125	1 05
Dallas	861	3 50
El Paso	1612	6 50
Galveston	800	Variable.
Greenville, Miss	292	1 50
Houston	710	3 40
Macon	257	1 60
Meridian	152	1 15
Mobile	2:8	1 90
Montgomery	96	1 10
Nashville	209	1 50
New Orleans	417	1 75
Pensacola	260	1 75
Savannah	448	1 80
Selma	101	1 20
Shreveport	473	2 25
Vicksburg	294	1 55

Remarks: Bunker rate to Mobile, $1.10; to New Orleans, $1.65; to Pensacola, $1.10. Export rate to Mobile, $1.10; to Pensacola, 1.05.

INDEX.

A.

	Page.
Alabama Iron and Steel Company	114, 119
Alabama Pipe Company	124
Alabama Rolling Mill Co.	119
Alabama Steel Works	120
Alice Furnace	10, 111
Anniston Bloomary—See Cherokee Iron Co.	
Anniston Pipe Works	123
Anniston Rolling Mills	120
Attalla Furnace	114

B.

Barton, A. E., on grading pig iron	136
Bay State Furnace	107
Bessemer Land and Improvement Co.	106, 122
Bessemer Ore	14, 15
Bessemer Rolling Mills	120
Bibb County, bloomaries in	10
Bibb Furnace	114
Birkinbine, John, ores of U. S.	16, 24, 27
Birmingham Rolling Mills	120
Birmingham Soil Pipe Co.	124
Blackband Iron Ore	13
Blair, A. A., Analysis of soft Red Ore	31
Bloomaries	123
Blue Billy	55
Bowron, James, on Pig Iron Market	126
Brannon, W. H., on Grading Pig Iron	143
Bridge Building Works	123
Brown Ore burdens	94, 104
" calcining	22, 52-54
" composition	48
" definition of	13
" improvement of	21, 22, 52, 54
" mining	46
" occurrence of	45
" phosphorus in	14, 48

	Page.
Brown Ore, price of	49, 92
" proportions of, in bank	47
" proportions used in furnace	14, 94, 104
" Russellville belt	7
, " screening, results from	51
" use of in charcoal furnaces	104
" used for car-wheel iron	9, 14
" used for pipe iron	14
" used in the state	13, 45
" valuation of	49
" variable nature of	14
" washing	46
"water in	17, 48
Buffalo Iron Co	114
Buchanan, Franklin, builds Tennessee	9
Burdens, furnace	78
Burdens, charcoal furnace	104
Burdens, coke furnace,	61, 62, 79, 84, 94

C.

Calhoun County, bloomaries in	10
Capital invested, in iron ore mining	3
Car Axle Works	124
Car Building Works	125
Car Wheel Works	124
Central Iron Works, See Shelby Rolling Mill Co.	
Charcoal, furnaces, list of	114
Chattanooga District, ores used in	1
Chattanooga Foundry and Pipe Works	123
Cherokee Iron Co	123
Clara Furnace	107
Clifton furnaces	107
Clinton formation, source of ore	29

162   GEOLOGICAL SURVEY OF ALABAMA.

Coals of Ala., Ga. and Tenn.. 1
Coal, for coking............... 77
" freight tariff. 158
" kind used for coking.. 78
" production of......... 158
" value of for coking.... 77
" yield of in coke....... 77
Cokes and Iron Ores, Southern....................... 1
Coke, analysis of.......... 70, 72
" analysis of ash of... 72, 73
" cell space of.......... 71
" character of coal used
    in making........... 78
" consumption of, in furnaces. 74–76, 84, 87, 94, 98
" cost of............... 92
" crushing, strain of.... 70
" first made in Alabama. 9
" freight tariff.......... 158
" furnaces, list of....... 106
" kinds of............... 68
" production of...... 77, 153
" specific gravity of..... 71
Colbert Iron Co.............. 107
Concentration of ore
   .... 1, 23, 37–39, 42
Coosa Furnace............. 115

D.

DeBardeleben, II. F 18
Decatur Charcoal Iron Furnace................. 115
Decatur Car Wheel and Manufacturing Co............. 124
Decatur Land Co............ 115
Davis-Colby kiln............ 52
D'Invilliers, E. V., on Southern Coke and Ore.......... 1
Dolomite, analysis of........ 58
" as flux, E. A. Uehling............. 64
" use of, largely increasing........... 58
" use of, due to C. A. Meissner........ 67
" valuation of....... 57

E.

Edwards Furnace............. 106
Elliott Car Co,.......... 124, 125
Engineering and Mining Journal, articles............... 1
Ensley Furnaces......... 89, 111
Eureka Coke Furnace....... 10

F.

Fleming, H. S., Ores used in Chattanooga District...... 1
Florence Cotton & Iron Co.. 108
Flue Cinder................. 55
Forges and Bloomaries...... 123
Fort Payne, furnace at...... 106
Fort Payne Rolling Mill, see Ala. Steel works.
Freight tariff—on coal 1896.. 158
" " on coke 1896.. 158
" " on pig iron 1888 136
" " " " 1893 141
'" " " 1896 148
Furnaces, charcoal, list of... 114
" coke, list of....... 106
" first in Alabama... 7
" progress of building, charcoal.... 118
" progress of building, coke........ 113

G.

Gadsden Foundry and Machine Works............... 124
Gadsden Iron Co............ 115
Gadsden-Alabama Furnace.. 107
Gholson coal seam, first coke
   in Ala. made from......... 9
Gordon, F. W., Large Furnaces on Ala. Materials 89
Gouge, definition of........ 30
Grading Pig Iron, agreement by Southern Iron Masters
   in 1888 141
Grading Pig iron, local practice...... 130, 132, 136, 143, 148

II.

Hard Red Ore—
behavior of in depth....... 41
calcination of 42
composition of.......... 36, 40
definition of.............. 29
effects of, on coke consumption..................... 87
effect of, on limestone consumption............. 84, 87
effect of, on quality of iron 90
occurrence of............. 39
price of................... 81
proportions used in furnace 84, 94
use of crushed........... 89
Hattie Ensley Furnace...... 107

INDEX

H

Hematite Ores—
- classification of 29
- iron chiefly made from 13
- occurrence of 29
- phosphorus in 15

Hematite Ores, see under Hard, and Soft Red.

Hercules Foundry 124
Hillman, T. T 10
Hood Machine Co 124
Hot Blast Stoves 119
Howard-Harrison Iron Co ... 123

I.

Iron Ores and Coals of Ala., Ga. and Tenn 1

Iron, Pig—
- freight tariff 136, 141, 148
- grading, see under Grading.
- production, charcoal 119
- " coke 114
- " cost of raw materials 84, 94, 104

Iron Trade Review——...... 27

J.

Jefferson Steel Co 121
Jenifer Furnace 115
Jones, Catesby ap. builds Tennessee 9

L.

Lake Ore 23, 26, 27
Laboratories, chemical 6, 14
Lady Ensley Furnace 108
Langdon Furnace 116

Limestone—
- analysis of 57
- consumption of 84, 87, 94, 98, 104
- compared with dolomite as flux 64
- cost of 92
- valuation of 57

M.

McCreath, A. S. on Southern Cokes and Ores 1
Magnetic Concentration 1, 22, 37–39
Mary Pratt Furnace 108
Meissner, C. A., use of dolomite due to 67
Montevallo, bloomary near .. 10
Morris, Geo. L 10
Michigan, iron ore production 26

Mill Cinder 55
Minnesota, iron ore production 26
Minnesota, value of ore in ... 16

O.

Ohio, pig iron production 23
Ore, Bessemer 14, 15
Ore, iron, analysis of, see under Brown, Hard, and Soft Red.
Ore, Mesabe 38

Ore, Iron—
- production of in Ala ... 25, 158
- sale of on analysis 19, 49
- semi-hard 40
- used in Chattanooga district 1
- varieties, see under Blackband, Brown, Hard, and Soft Red.
- value of in Ala. and United States 16, 25, 28

Oxmoor Furnace 10, 110

P.

Peacock's Iron Works .. 124, 125
Pechin, E. C. 1
Pennsylvania, iron ore production 26
Perry, Matthew Calbraith ... 9
Philadelphia Furnace 108
Phosphorus in Ala. Ore 15
Piedmont Land & Improvement Co 116

Pig Iron—
- change of nomenclature in 1888 141
- charcoal, production of 119, 158
- coke, production of 114, 158
- cost of making 5
- cost of raw material in making 84, 94, 104
- freight tariff, 1888 136
- " " 1893 141
- " " 1896 148
- grading .. 130, 132, 136, 143, 148
- grades effected by burden 84, 90, 94
- market 126
- prices of in 1888 135
- " " " 1893 141
- " " " 1896 147
- production—charcoal .. 119, 158
- " —coke 114, 158
- " —total yearly .. 158

	Page.
Pioneer Furnaces	108
Pipe Works	123
Polksville, charcoal furnace at	8
Porter, Jno. B., Iron Ores and Coals of Ala., Ga. and Tenn.	1
Pratt Coal Mines, paper on by E. Ramsay	1
Puddle Cinder	55
Purple Ore	55

R.

Ramsay, Erskine, on Pratt Coal Mines	1
Residue from Acid Works	55
Robertson, Kenneth, on grading pig iron	132
Rock Run Furnace	116
Rolling Mills in Alabama	119
Round Mountain, charcoal furnace at	8
Round Mountain Furnace	116
Russellville Brown Ore	7

S.

Schultz, Captain	10
Selma, Confederate arsenal at	9
Sheffield Furnaces	109
Sheffield, ore used at	7
Shelby, charcoal furnace at	8, 117
Shelby Rolling Mill Co	121
" Iron & Steel Co	109
Sloss, J, W	10
Smith, Eugene A	1, 9, 10
Soft Red Ore—	
composition of	31, 32
concentration of	37, 38
definition of	28
exhaustion of	37
improvement of	22, 38
occurrence of	30

	Page.
Soft Red Ore—	
price of	5, 1, 928
physical nature of	34
proportion of used in furnaces	84, 94
Soil Pipe Co., Birmingham	124
Southern Bridge Co	123
Southern Cokes and Iron Ores	1
Spathite Furnace	109
Spathite ore	23
Stoves, Hot Blast, in Ala	119
Swank, Jas. M	7

T.

Talladega Co., bloomary in	10
Talladega Furnace	110
Tap Cinder	55
Tecumseh Furnace	117
Tennessee, iron clad ram	9
Tennessee Coal, Iron & Ry. Co	110, 111, 122
Texas, value of ore in	16
Thomas, Robt	8
Trussville Furnace	111

U.

Uehling, E. A., use of dolomite as flux	64
Union Iron Works	125
United States Car Co.	121, 124, 125

V.

Valuation of Ore	17, *et seq*.

W.

Ware, Horace	8
Williamson Furnace	112
Witherby, E. T	8
Woodstock Furnaces	112, 117
Woodward Iron Co	112

www.ingramcontent.com/pod-product-compliance
Lightning Source LLC
Chambersburg PA
CBHW031454160426
43195CB00010BB/974